made. In addition to numerous explanatory diagrams, there are also five do-it-yourself projects for making a sextant; a steelyard (a Roman hand balance); an astrolabe; a sundial; and a pendulum clock.

Owen Bishop taught biology, and then worked for the United Nations as a science and education advisor, and a developer of science curricula. He is the author of several books, and lives in England.

# YARDSTICKS OF THE UNIVERSE

# YARDSTICKS
# OF THE UNIVERSE

OWEN BISHOP

**Peter Bedrick Books**

New York

First American edition published in 1984 by
Peter Bedrick Books
125 East 23 Street
New York, N.Y. 10010

Published by agreement with Frederick Muller Limited, London

Library of Congress Cataloging in Publication Data
Bishop, Owen.
　　Yardsticks of the universe.
　　Includes Index.
　　1. Physical measurements. I. Title.
QC39.B57　　　1984　　　530.8　　　83-15782
ISBN 0-911745-17-3 hardcover
ISBN 0-911745-42-4 paperback

Manufactured in the United States of America
Distributed in the USA by Harper & Row
and in Canada by Book Center, Montreal

# Contents

**Chapter 1**

# Yardsticks

People have been using yardsticks for over a thousand years (Fig. 1). The word 'yard' comes from the Anglo-Saxon word 'gyrd' or 'gierd' which means 'a stick'. In those days a stick that was one yard long was a handy tool for measuring a plot of land, a wall being built or a length of cloth. Even today you may find a yardstick being used in a shop that sells fabrics.

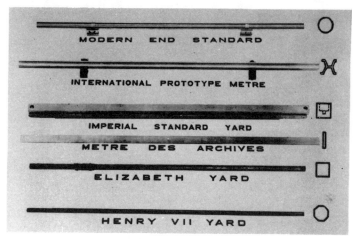

**Fig. 1.** Some historical and modern yard and metre standards. *Photo by Crown Copyright, National Physical Laboratory.*

In the early days yardsticks were not all exactly the same length. There was no *standard yardstick* against which all other yardsticks could be measured. This caused many difficulties. It made it easy for dishonest tradesmen to cheat their customers. Then King Edgar, the Anglo-Saxon ruler of England from 957 to 975, decided to set up a standard yard. He defined this as the 'measure of Winchester'. Probably this description referred to one particular yardstick carefully kept at Winchester, which in those days was the capital of England. In doing this he did two important things.

1) He decided which one of many similar but differing lengths would be called 'a yard'.

2) He set aside one actual yardstick as a standard.

From that time onward there could be no argument about the length of a yard. In those days only someone such as a king had enough power to set up such a standard and be sure that others would accept it. Nowadays such standards are decided on by groups of people appointed by governments or by international organisations.

King Edward was not the first person to set up a standard. Some units and standards from earlier times were based on common everyday objects and experiences. This is because measurement is important to almost everyone in everyday life. It is especially important to traders and farmers. Measurement is a *practical* activity and people express measurements in practical ways. For example, if someone asks you how far it is from your house to your school, you would probably say, 'Its about ten minutes walk' or 'It takes half-an-hour on the bus'. You probably have no idea of the exact distance in yards or metres. Instead you express the distance in practical terms — *how long* it takes to journey from home to school — for this is what matters most. You use time as a yardstick for distance.

Many of the early measurements, such as the stadion (the distance run without getting out of breath — about 200m), the furlong (the distance a horse can pull a plough

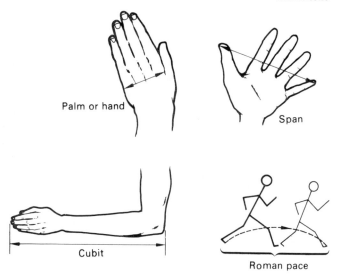

Palm or hand

Span

Cubit

Roman pace

**Fig. 2.** Units based on the human body. A yard may be a 'double cubit'.

without stopping for a rest) and the acre (the amount of land two yoked oxen can plough in one day) were based on practical activities. Such measurements were not exact, and would easily vary from one country to another, or even from one village to another — but they had the advantage that they were easy to understand and easy to reproduce. Measurements based on the *palm* used something that is always ready to be used — the human hand. Today the *palm* is still used for measuring the height of horses, though it is now called a *hand*. It is defined as a length of exactly 4 inches. People still use it partly because it is a convenient size for measuring horses and partly because of custom. Also, if there is no tape measure available, you can always measure the height roughly by using your hand. Similarly the *foot* is based on the size of an average man's foot.

Seeds were often used as standards for weight. One such weight was the grain — until recently the smallest unit of weight of the British system (1/7000 of a pound, or 65mg). It was used by pharmacists when weighing out small amounts of substances for making medicines. Another seed weight was the carat, which is the weight of a seed of the carob tree. A *small* daisy plant, growing from a crack in the pavement produces one or two daisy flowers of standard size, not lots of small-sized flowers. This feature of plants, that their flowers, fruits and seeds tend to be of standard size no matter where the plant is growing, makes fruits or seeds very suitable as standard weights. Provided that mature, disease-free seeds are used, we have a ready supply of standard weights from the field or forest. If weights are lost or damaged, it is simple to obtain new ones that will be similar to the old ones. Larger or smaller units of weight can be based on these naturally occurring weights.

Another useful practice is to relate the units of one quantity to units of another quantity. In the earliest days a special jar or bucket would be used as a standard for measuring volumes. It would be used for measuring anything from corn to wine. One such jar would be kept safely as a standard for all other measuring jars. If the standard was damaged or lost, there would be problems in making another one exactly like it. Also, as trade expanded across the world, there would be a need to be able to make lots of similar jars to be used as standards in each country or town. It would not be practicable to take the standard jar from place to place to be copied. The Babylonians had a system of related units that made it possible to construct a standard of volume and weight whenever needed. Their unit of length was the *palm*. A cubical container with sides measuring 1 palm had a volume known as a *ka*. This is approximately 1 litre by modern standards. Given the breadth of the human hand, a skilled man could easily make a box with a volume of 1 ka. The unit of weight was in turn related to the ka.

The unit called the *great mina* was the weight of water required to fill a container of capacity 1 ka. Given a container, some water and a balance, a man could then make a stone or metal weight of 1 great mina. Using this weight, other weights could be made according to the system:

| | | |
|---|---|---|
| 1 shekel | = | 1/60 great mina |
| 1 grain | = | 1/60 shekel |
| 1 talent | = | 60 great mina |

A man, given the necessary materials, and knowing the relationships, could make a usable set of weights and measures of capacity anywhere in the world. As will be described later, the International System of Weights and Measures (called SI for short) that we use today also has related units. There are only seven base units which have specially defined standards and all other units in the system are related to these seven.

Our modern system of counting is based on the number 10. We call it the decimal system. This probably began when early men used their fingers and thumbs for counting small numbers of objects. In other ages and places, people used other number systems. The Babylonians used a system based on 60. It was natural for them to divide the great mina into 60 shekels, and to put 60 great minas together to make 1 talent. This system survives today in our measurements of time (60 seconds = 1 minute; 60 minutes = 1 hour) and angle (60 seconds = 1 minute; 60 minutes = 1 degree). It would cause so much confusion to change our units for time that the second, minute, hour, day and year have been accepted as part of SI, which is otherwise strictly decimal. In some systems the divisions have been into 12. For example, 12 inches = 1 foot, or 12 furlongs = 1 league, or 12 pence = 1 shilling. The quantity twelve (a dozen) can readily be divided into halves (6), thirds (4), quarters (3) and sixths (2). This is very convenient in trade and agriculture, though very inconvenient for scientists and mathematicians working in the decimal system.

In more modern systems there has been a trend to decimal division. The first major decimal system was the *metric system*. This was devised in France at the end of the 18th Century. Its base unit was the metre, divided into 10 decimetres, which were each divided into 10 centimetres and again into 10 millimetres. For longer distances, 10 metres made 1 decametre, 10 decametres made 1 hectometre and 10 hectometres made 1 kilometre. The metre, millimetre and kilometre have been adopted into SI, and the centimetre is still in common use, but the other metric units of length are seldom used. From base unit, the metre, the unit of area was derived. This is the *are*, equal to 100 square metres (10 × 10). As a measure of land area we generally use the *hectare* ( = 100 ares = 10,000 square metres). As a measure of volume the metric system used a cube measuring 1/10m × 1/10m × 1/10m and called this a *litre*. A cubic centimetre of distilled water at 0°C gave the unit of weight, the *gram*.

With the increase in international trade, the large number of local standards of measurement were gradually replaced by those few that were more widely known and used. By the beginning of the 20th century the British or Imperial System and the Metric System had become the two main international systems. Their use brought great advantages to trade between nations. It also made possible an increasing amount of co-operation between scientists working in different countries. In the engineering and construction industries the making of girders, bricks, nuts, bolts and all kinds of materials and parts in standard Imperial or Metric sizes meant that, when a machine or a bridge was being built, its parts would fit together accurately. If spares or replacements were needed, they too would fit comfortably into place. Instead of each piece having to be made by hand to fit in place, mass-production became possible. This is a further benefit of having a standard system of yardsticks.

The metric system was so successful that its use has spread almost throughout the whole world for use in scientific measurements as well as in trade and everyday life. Much of it has been used as the basis of SI. Scientists have been able to use the metric system to devise related units. Examples are the units for energy (*joule*), power (*watt*), and force (*newton*). These and many more are our yardsticks for the universe.

**Chapter 2**

# Bettering our Senses

The sense organs of the human body tell us what is happening in the world around us. They help us see how big things are, or feel how heavy they are, or tell how hot they are. We use our senses a lot for measuring. It is quicker and more convenient to use senses than to use a ruler or a balance. But just how reliable are our senses? With practise, we can learn to judge distances fairly precisely, though not with the precision of a millimetre scale. With practise, we can learn to estimate mass, though not with the precision of a balance. If we are not very experienced at estimating the sizes or masses of objects in millimetres or grams, we can at least compare two objects and say which is bigger or which is more massive. This sounds simple enough, but there can be unexpected sources of error. Cut out two paper shapes like those of Fig. 3 making them *exactly* alike in size and shape. Lay them side by side. Which one looks bigger? Now swap them round so that the shape that was on the left is now on the right. Which one looks bigger now? This optical illusion shows that our senses can be deceived without our being aware of it. If we were comparing two objects shaped like the pieces of paper, we would say that one was bigger than the other. Of course, if we place the paper shapes together, we can see that one matches exactly against the other. We then know that they are

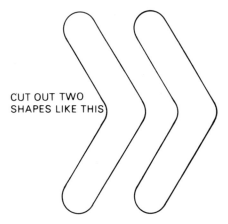

CUT OUT TWO
SHAPES LIKE THIS

**Fig. 3.** An optical illusion.

both the same size. When we are comparing sizes of other objects by eye, it may not be possible to bring them together to match them. The accuracy of our estimate may be affected in various unexpected ways by one or more optical illusions. The only reliable method of measuring is to use a scale or some similar instrument.

The nerve endings in the skin that respond to heat and cold are even less reliable than the eye. If you have three bowls of water (Fig. 4), you can demonstrate just how unreliable our heat sense is. After placing your hands in hot water (as hot as you can comfortably stand) and cold water for a few minutes, place *both* hands in the middle bowl. To the hand that was in the hot water, the water in the middle bowl feels *cold*. To the hand that was in the cold water, the water in the middle bowl feels *hot!* The sense endings of the skin may be useful to warn us when we are touching something that is dangerously hot or cold, but they are of very little use for judging temperature. The only reliable method is to use a thermometer or some similar instrument.

Our senses are certainly not precise enough, accurate enough or reliable enough to use for measurements in

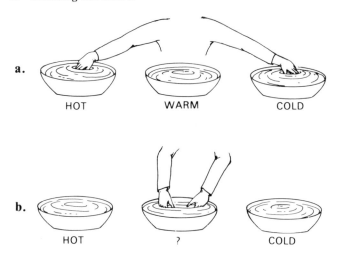

**Fig. 4.** Another illusion.

science. We need to use instruments. These not only give us high accuracy and precision but may also be able to measure beyond the range of our senses. To the sense organs of the skin, all temperatures above about 50°C are simply 'too hot to touch'. They all produce the same extreme sensation. Temperatures above 50°C are beyond the range in which we can attempt to judge different levels of hotness or coolness. Similarly, a micrometer screw gauge allows us to measure lengths to a hundredth of a millimetre, a distance that we can not estimate by eye alone. Instruments also are used to measure things our senses can not detect, such as the radio waves from distant stars or the ultra-sounds from flying bats.

*Measuring temperature*
It is important for scientists to be able to measure temperature. Temperature affects the properties and behaviour of objects and substances in many ways. An increase in temperature may cause a substance to decompose or explode. Most substances expand as their temperature increases and contract when it decreases. A

living organism is much affected by the temperature of
its surroundings. Small changes of temperature can produce large effects. This is why our senses are of little use
for estimating temperature in scientific work. It is only
by inventing instruments that measure temperature, that
scientists have been able to investigate the many effects
that temperature has on the physical and chemical properties of matter.

One of the first people to invent a thermometer was
Galileo. Toward the end of the 16th century he invented
the thermoscope (Fig. 5). The air inside this expanded as
temperature increased and pushed the coloured water
down. As the temperature fell, the air contracted and
sucked the coloured water up the tube. A scale beside the
tube was marked with divisions to indicate the

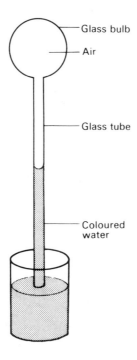

Glass bulb

Air

Glass tube

Coloured
water

**Fig. 5.** Galileo's thermoscope.

temperature. The thermo*scope* was given this name because it made it possible for the observer to *see* when temperature changes were happening. Air expands a lot when heated; with a large bulb and narrow tube the level of water would change a lot with only a small change of temperature. As an instrument for detecting small changes of temperature, the thermoscope was precise. Unfortunately it was not accurate, for it was also affected by changes of atmospheric pressure. When atmospheric pressure increased, the water would be forced further up the tube. The thermoscope would give a temperature reading lower than the true one. The readings would also be affected by the amount of water in the bowl. In time, some of this would evaporate, leading to an inaccurate reading. Lack of accuracy meant that readings taken on one day would not be compared with readings taken on another day. In spite of this, the thermoscope was found to be far more reliable than the human senses.

The next step was to invent an instrument that was completely sealed. This was a true thermometer, invented in the mid-seventeenth century. The liquid used was alcohol. Since the instrument was completely sealed it was not affected by changes of atmospheric pressure. Also the alcohol could not evaporate. For these reasons it was a much more accurate instrument than Galileo's thermoscope. It could be used for comparing readings taken on different occasions.

The final step toward the thermometer that we use today was taken by Newton at the beginning of the eighteenth century. Until that time the scales of thermometers had been marked in various ways by their makers, but there was no standard scale used by everybody. Newton proposed a scale of degrees, running from 0° as the temperature of freezing water to 12° as the body temperature of a healthy person. Here were two standard temperatures, or *fixed points* that could be used by anyone wanting to calibrate a thermometer. Wherever this scale was used, a given number of degrees would always have the same meaning. Different scientists, working in different places with different thermometers, but both using Newton's scale, would now be able to understand each others results. Following the invention of this thermometer with its scale based on two fixed points, big scientific advances were soon made in the study of heat. It became possible to measure the temperature changes when heat passes from one body to another. This led to the ideas of specific heat capacity and specific latent heat, ideas that play a big part in the science of heat today.

Newton's scale of temperature was soon replaced by another scale, invented by Fahrenheit. To fix the lowest point on the scale (0°F) Fahrenheit used a freezing mixture. This is a mixture of ice and salt and has a temperature lower than that of melting pure ice. The mixture used to fix 0°F had the coldest one known. To fix the highest point he used the human body temperature, and called this 100°F. His scale had a hundred degrees while Newton's had only twelve. Although Fahrenheit's scale ran over a slightly greater range than Newton's, its degrees represented a much smaller temperature step, so it was easier to record temperatures precisely as whole numbers, without using fractions of a degree. Fahrenheit's other development was to use mercury in the thermometer instead of alcohol. This gave added precision to measurements.

The Fahrenheit scale remained in use for over 200 years, though its fixed points were later re-defined. It is not possible to fix the lower point with high precision by using a freezing mixture, for much depends on exactly how the mixture is made. A new lower fixed point was used, the melting point of pure ice. This was defined as 32°F. At the other end of the scale another new standard was used, the boiling point of water at a pressure of 1 atmosphere. This was defined as 212°F. This could be fixed with far greater precision than the point based on body temperature. The temperature of a healthy person is 98.6°F, not 100°F as was first defined. In making his original scale Fahrenheit had measured the temperatures of some hospital patients, and it is thought that these may have had a slight fever at the time.

Another temperature scale was devised by Celsius, using the melting point of ice and the boiling point of water as fixed points. The lower point was defined as 0°C. Since this scale had exactly 100 degrees, it was called the Centigrade scale. It is still referred to by that name nowadays, though officially it is known as the Celsius scale. In recent years it has replaced the Fahrenheit scale in almost all countries in the world. The International System uses the kelvin as its unit of temperature. This has its zero at the coldest temperature possible (absolute zero, 0 K or −273.15°C). One kelvin represents the same temperature *difference* as one degree on the Celsius scale, so 0°C is 273.15 K and 100°C is 373.15 K. This scale is used in scientific work.

To measure temperature we generally use a physical property that changes uniformly as temperature changes. In Galileo's thermoscope and the mercury and alcohol thermometers, we use the fact that substances expand when heated and contract when they are cooled. Another property that can be used is the change in the electrical resistance of metal. A platinum resistance thermometer is very accurate and can be used for measuring temperatures higher than those measured by a mercury thermometer. A thermocouple can measure temperatures even higher, up to about 500°C. Since a thermocouple is small in size it can be used to measure temperatures of small objects or regions. If the temperature is changing rapidly the thermocouple, being small, changes temperature rapidly too. This makes it especially useful for measuring temperatures that are changing quickly. The thermistor is similar to the platinum resistance thermometer, since it depends upon change of resistance with temperature. The difference is that the thermistor is made of semi-conducting material (similar to the materials used in making transistors) and its resistance changes a lot for a small change in temperature. The instrument can measure temperatures of 0.01 degree. A thermistor can be a very small bead of material, so it can be used in small spaces and to measure rapidly changing temperatures.

*Too hot to touch*
The thermocouple and platinum resistance thermometer can operate at high temperatures. This is one way in which they are better than the human senses. If we are faced with the problem of finding the temperature of something that is too hot to touch, the best we can do, without using instruments, is to hold out our hands toward it. We feel the radiant heat that strikes our palm. Several temperature measuring instruments called *pyrometers* are based on the idea of measuring the radiation given off by hot objects. A simpler and more direct method that is sometimes used to measure temperatures of furnaces or pottery kilns is to put pyrometric cones in the furnace. These are made of ceramic materials and are made from different mixtures that melt at different fixed temperatures. If one cone melts and the one with the next higher melting-point does not melt, we know that the temperature of the furnace lies between the melting-points of the two cones. A similar method is to use paints that change colour at set temperatures. Such methods are simple and cheap but are not precise.

*Infra-red detectors*
Our eyes are not sensitive to infra-red, which is the radiation just beyond the red end of the spectrum. Our skin can detect it as a warm feeling, for instance, when we feel the warmth from a coal fire. Some animals are able to detect infra-red radiation better than we can. Owls hunting at night can·sense the body-heat of small mammals such as mice. The pit-viper has groups of infra-red detectors in pits beside its eyes and uses them for hunting in the dark.
Infra-red radiation can also be detected by devices such as photoelectric cells and phototransistors. These

can be placed at the focal point of large reflecting telescopes. In this way astronomers look for stars that are emitting infra-red radiation. The stars that we can see by eye (with or without a telescope) are hot enough to emit both infra-red and visible radiation, but there are also many cooler stars that emit little or no visible light. These are invisible to the eye, even with a telescope. By using an infra-red detector in a telescope we are able to study these infra-red stars and measure their temperatures. In this way several thousands of previously unknown stars have been discovered.

Returning to Earth, we also use electronic infra-red detectors in a variety of ways to measure temperature. One of these is the *thermal scanner* (Fig. 6). The scanner displays an outline of the object before it. The image consists of small squares and the colour of each square indicates the temperature of the object at that point. The scanner is sensitive to very small differences in temperture. If the human body is scanned, we get a picture showing the temperature of the skin at all points. Such information is very useful for studying the way heat is brought to the surface of the body by the blood. It helps us find out about the workings of the blood system and may help in tracing disorders of the body.

*Measuring brightness*
Brightness, or intensity of light is something that our eyes seem good at sensing. We can tell which is the brighter of two lamps or which is the lighter of two areas of paper. Or can we? Look at Fig. 7. Which of the two grey areas is the lighter? Actually they are both exactly equal in brightness, but the fact that one is surrounded by black makes it look brighter than the one surrounded by white. You may also think that the grey area surrounded by black looks *smaller*, though both areas are the same size. Another difficulty in using our eyes to estimate brightness is that they slowly adapt to levels of light. If you visit the cinema on a dull rainy day, the sky looks very bright as you come out. Yet after a few

**Fig. 6.** The heat pattern produced on the screen of a thermal scanner. This one is of an eight-year-old boy with his Airedale dog. Note the healthy, cold nose. *Photo copyright 1980 by Dr. R. P. Clark and M. R. Goff.*

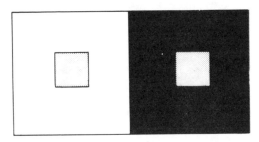

**Fig. 7.** Which grey area is the lighter?

minutes the effect is gone and the sky looks its true dull grey. As we found before, to measure accurately we must use an instrument. An instrument for measuring light is called a *photometer*. An early and simple photometer was invented by Bunsen in 1844. This was his grease-spot photometer. You can easily make one for yourself by putting a drop of cooking oil on a piece of paper (Fig. 8).

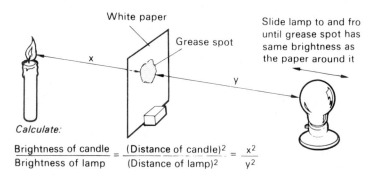

**Fig. 8.** Using a grease-spot photometer to compare brightness.

Nowadays we nearly always use electrical methods for measuring brightness. Photographers need to measure brightness so that their photographs are correctly exposed. There are several different ways in which light can be detected and its intensity measured. All of these are made use of in photographic exposure meters of different kinds.

The early astronomers noticed that the stars are of different brightnesses. Most of them are constant in brightness, though some are variable, as we shall see in Chapter 5. Hipparchus, one of the greatest astronomers (2nd Century B.C.), first invented a way of describing the brightness of stars as seen by the eye. He gave us the system of star *magnitudes* that we still use today. The brightest stars of all he called 'first magnitude stars'. Those that could only just be seen were of sixth magnitude. He placed the other stars in four groups, magnitudes 2 to 5. When modern astronomers measured the brightness of stars by attaching a photometer to a telescope, it was found that first magnitude stars were about 100 times brighter than stars of the sixth magnitude. The brightness of stars is now measured precisely by photometer and they are given their magnitude on a scale on which six magnitudes equal a difference in brightness of exactly 100 times. Using telescopes, we can detect and measure brightnesses much less than sixth magnitude. We can also measure them more precisely, to a tenth of a magnitude or less. Precision of this kind is impossible by eye alone.

Another measuring instrument that measures brightness is the colorimeter. In the next section we shall see one way in which this is used to better our senses.

### Acid or Alkali?

Our sense of taste is one way of detecting the presence of acid. Acids have a sharp, sour taste. The sharpness of citric acid in lemon-juice or the malic acid in unripe apple or the lactic acid in sour milk are familiar to us all. The taste-buds of our tongue are sensitive to relatively small quantities of weak acids. In small quantities the carbonic acid in carbonated fizzy drinks produces a pleasantly refreshing effect. As we have found before, our senses are not reliable enough for scientific use. Acids frequently used by chemists are too strong for our taste buds. They damage the tissues and many are poisonous too. Amounts of particular acids can be measured by various chemical tests. If we wish to measure the level of acidity in general, we use a pH meter. Before describing how this instrument works, we must be more exact about what it measures. In pure water a proportion of the molecules ($H_2O$) split to form ions. There are $H^+$ (hydrogen ion, with a positive electric charge, having lost an electron) and $OH^-$ (hydroxyl ion, with negative electric charge, having gained an electron) ions. We say that water and any other solution in which $H^+$ and $OH^-$ ions are present in equal amounts are *neutral*. In acidic solution, there are more $H^+$ ions than $OH^-$: in an alkaline solution (for example a solution of an alkaline substance such as sodium hydroxide, NaOH) there are more $OH^-$ ions than $H^+$ ions. If we can measure the concentration of hydrogen ions present in a solution, we can tell how acidic or alkaline it is. This is expressed on the pH scale. A value on the pH scale is related to the concentration of hydrogen ions. On this scale neutral is represented by pH7. Acid solutions, with high hydrogen-ion concentrations, have values below 7, down to about pH1; alkaline solutions have values above 7, up to about pH13. Certain coloured substances which we call indicators have the property of changing colour according to the pH of the solution they are mixed with. A well-known example is litmus, which changes from red in moderately acid pH (pH5) to blue in moderately alkaline pH (pH8). Another indicator is phenolphthalein, which is colourless in acid and neutral solutions and gradually turns red between pH8.3 and pH10. Mixtures of indicators can be prepared so that they have a range of colours according to pH. Their colour is matched by eye against a series of colour standards, so that it is possible to measure pH with the precision of 0.1 on the scale. It is not easy to match colours precisely by eye, particularly if the solution itself is coloured. Colour-blindness may upset the ability of a person to match colours correctly. People can be colour-blind without being aware of it and their measurements are inaccurate. To avoid such errors we can use a colorimeter. We first prepare a series of standard solutions of known pH and add indicator to these. For example, we may have solutions of pH3, pH3.2, pH3.4, pH3.6 and pH3.8. Each of these causes the indicator to become a slightly *different* colour. Each standard solution is placed in the colorimeter and the brightness of the light passing through it is measured. We plot a graph of brightness (the meter reading) against pH. Now we put some of the same indicator in a solution of *unknown* pH. This is placed in the colorimeter and the meter reading is taken. Using our graph we can find out the value of pH that corresponds to this reading. This is an example of how we use a series of known standards to *calibrate* a measuring instrument.

Higher precision, to within 0.01 on the pH scale, is obtained by using an electronic pH meter. This measures the electrical potential due to the hydrogen ions in the solutions. The electronic circuit measures this potential and indicates the pH on a meter.

In this chapter we have seen how the invention of measuring instruments has made it possible for us to improve upon the accuracy and sensitivity of our senses. We have also been able to extend the range of such measurements far beyond the range that our unaided senses can respond to. There are many more such examples of this throughout the book. Using the human sense organs alone, early scientists first began to study the world around them. But their progress was greatly limited by the nature of the human senses. As we shall see in the next chapter the invention of measuring instruments, even quite simple ones, made countless new discoveries possible.

**Chapter 3**

# Measuring the Earth

It is easy to measure the length of an object when the scale we use is longer than the object we are measuring. It is much more difficult to measure an object that is many times bigger than our biggest scale. Before we can do this we need to invent special techniques. To measure great distances we need techniques based on geometry. The name 'geometry' itself means 'Earth-measuring' for this is why the ancient Egyptians first began to study the properties of triangles and other geometrical figures. Their reasons were practical ones — the construction of buildings and the measuring and marking out of areas of land after the annual floods of the Nile. Thales, a Greek philosopher of the 7th century B.C., extended the use of geometry in several ways. He devised a method of measuring the distance of ships at sea (Fig. 9). The actual *measurement* was then on a scale of convenient size. From this result, a *calculation*, based on the idea of similar triangles, gave the distance of the ship. Such techniques were used for measuring distances on land and sea and were the basis for the early maps. In about 200 B.C. the first attempt was made to measure the circumference of the whole Earth. It had been believed for several hundred years that the Earth was a sphere, so it was natural to want to know its circumference. The method used by Eratosthenes was based on simple

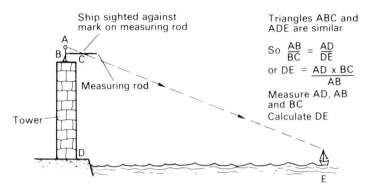

Ship sighted against
mark on measuring rod

Triangles ABC and
ADE are similar

So $\frac{AB}{BC} = \frac{AD}{DE}$

or DE $= \frac{AD \times BC}{AB}$

Measuring rod

Measure AD, AB
and BC

Tower

Calculate DE

**Fig. 9.** Thales method of measuring the distance of ships at sea.

geometrical ideas. He noted that at midday on June 21, midsummer day, the Sun was vertically overhead at Syene, a town in Egypt. At Syene there was a deep vertical well and the suns rays shone directly down the well, to the very bottom. In Alexandria, where Eratosthenes was librarian, the Sun was not vertically overhead at that time and date. He used a scaph (Fig. 10) to measure the angle between the Sun's rays and the vertical direction. He found the angle to be one fiftieth of a circle ($= 7.2°$). He knew that Alexandria was 5000 stadia (p.2) due north of Syene and so he was able to calculate the diameter of the Earth (Fig. 11). His reasoning was that if one fiftieth of a circle is equivalent to 5000 stadia, then a whole circle is equivalent to 50 × 5000 stadia, that is 25000 stadia. The stadion was about 185m so in modern units his result gives the circumference of the Earth as 46250 km. This is very near to the value obtained by modern measurements, which give about 40000 km, depending on exactly over which part of the Earth the measurement is made. Eratosthenes obtained a surprisingly accurate result, considering that his instruments were so simple. However, Alexandria is not precisely due north of Syene, and the angle of shadows cast by the Sun at Alexandria is actually greater than $7.2°$. Luckily the

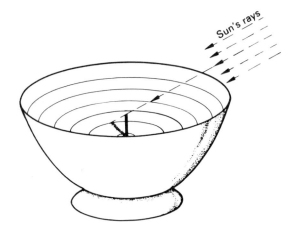

**Fig. 10.** A scaph, used for measuring the angle of the sun's rays. The shadow of the stick falls on the interior of the scaph. The angle is measured by seeing which line the tip of the shadow touches.

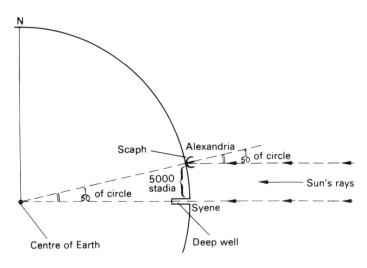

**Fig. 11.** Eratosthenes' measurements.

errors in Eratosthenes' measurements tended to cancel out! A method very similar to that used by Eratosthenes was used by Jean Picard in France in the 17th century. By that time, telescopes had been invented so measurements could be made with greater precision. Instead of using the Sun, he used a star and measured the angle between the vertical direction in two places that were a known distance apart. His result, gave the circumference of the Earth as 40040 km, a distance in good agreement with modern figures. Soon after this, investigations showed that the Earth was not a perfect sphere, as people had originally thought. On an expedition to Cayenne, French Guiana, it was discovered by the French astronomer Jean Richer, that pendulum clocks ran more slowly in Cayenne than in Paris. Although his clock had been set very accurately in Paris, it lost about 2.5 minutes a day when in Cayenne. He explained this by suggesting that the force of gravity was less at Cayenne. Cayenne is near to the equator and it was suggested that the force of gravity there was less than at Paris because the Earth bulged at the equator. Measurements of the same kind have since shown that this is true. By using such methods we have been able to measure the diameter of the Earth in various directions. It has been found that its diameter from north-pole to south-pole is about 43 km less than its diameter at the equator. In other words, it is flattened at the poles and bulges at the equator. More recent measurements have shown that the shape is even more complicated than had been supposed. When an artificial satellite is on orbit around the Earth, its motion is strongly affected by the Earth's gravitational field. This depends on the shape of the Earth. We have been able to make high-precision measurements of the orbits of satellites and then to calculate the shape of the Earth from these. Two English scientists, Cook and King-Hele measured the orbit of two Russian *Cosmos* satellites over a period of several years. Their calculation, accurate to half a metre, show us the true pear-like shape of the Earth. This shape is called a *geoid*.

*Mapping the Earth*

Many of the surveying instruments used today have been in use since the time of the Romans. Although the precision of the instruments is greater now, the ways in which they are used have remained unchanged. The basic principle is that of triangulation. A base line is measured with the greatest accuracy possible. Then bearings are taken from either end of the base line to fix the position of a third point. The length of all three sides of the triangle are now known accurately. All three sides may now be used in base lines for further triangles. In this way a system of triangles can be set up to cover a whole country. This serves as a framework for detailed mapping. The triangulation method depends on our being able to see from one end of the base-line to the other and also to the third point of the triangle. Hills or other high spots are frequently chosen as triangulation points, so that the triangles can cover as large an area as possible. The curvature of the Earth sets an upper limit to the size of the triangle. For triangulation that is to cover larger areas, a method known as flare triangulation is used. In Fig. 12 we see two accurately mapped triangles that are so far apart that it is not possible to see any part of one from any part of the other. Perhaps there is sea between them or a high and inaccessible mountain range. A flare is carried up to great height by a balloon or rocket, so that it can be seen from both areas. The flare is sighted and angles are measured from the points of each triangle. The relative positions of the two triangles can then be calculated and they can be plotted on one map. The maps of Norway and Scotland have been placed in correct relation to each other by this method, with an accuracy of 2 or 3 metres. Instead of using a flare, an artificial satellite can be sighted from each area. It is also possible to use our natural satellite, the Moon. In these measurements and also in most modern methods of triangulation we do not measure the *angles* of the triangles, but the *lengths* of their sides. The tellurometer projects a beam of very-high-frequency radio waves (microwaves or radar) from

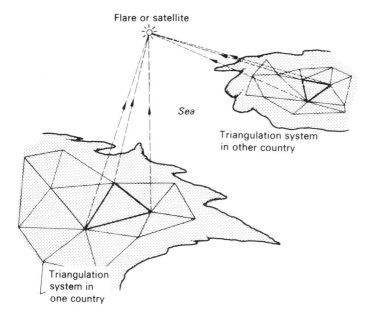

**Fig. 12.** Flare triangulation.

one corner of the triangle to a reflector placed at another corner. Or there may be a second transmitter at the far end that sends back a beam of microwaves as soon as it receives the beam from the first transmitter. The tellurometer is used to measure how many wavelengths and parts of wavelengths of the radio transmission fit into the side of the triangle. The length of the side can then be calculated. This instrument and the Geodimeter, a similar instrument using beams of light, can measure distances of up to 30 km with a precision of a few millimetres. After a triangulation survey, details can be filled in by local surveys. Nowadays, we use photographs taken from aeroplanes or from satellites. The Landsat satellites survey the whole surface of the Earth every 18 days. They make 14 north-to-south orbits each day, scanning the Earth beneath them with instruments sensitive to different wavelengths of visible and infra-red radiation.

Their function is not the mapping of towns, coastlines, mountains or other main geographical features, but the mapping of the resources of the Earth. They can map the distribution of forests and crops. They can measure the amounts of heat being lost from different regions of the Earth. They detect regions of pollution of air and water. Even from their great height (just over 900 km) they can detect objects as small as a house. They can help us investigate the geology of remote areas and discover new places in which to mine for rare and important minerals.

The early explorers had no such assistance from technology. Their lot was to put to sea and sail away into the unknown. If they were not able to find their way back, they might well perish from shipwreck on unfamiliar coasts, or from the disease of scurvy, the result of lack of vitamin C in their diet. The early sailors and navigators tried to chart their courses and produce maps that others could use. Their biggest problem was to know *exactly* where they were at any given time. One of the instruments they used was the sextant (see Make a Sextant, p.30). It is used to measure the angle between the Sun and the horizon. If this is measured at local noon, when the Sun is at its greatest height in the sky, it is easy to work out the latitude of the place at which the measurement is made (Fig. 13). The explorers and sailors had no difficulty in finding their latitude, but in order to chart their exact position they also needed to know their longitude. The method of finding longitude depends on the fact that for every degree west you travel, your local noon is 4 minutes later than the local noon of the place from which you started your journey. For example if you are 2°W of Greenwich, your local noon is 8 minutes later than noon GMT. If you travel east, local noon is earlier. To measure how many degrees of longitude you have travelled you must know what the time is in the place from where you began your journey (or at a standard location such as Greenwich). In those early days, voyages lasted several months and there were no radio signals to indicate GMT. At the equator a degree of longitude is

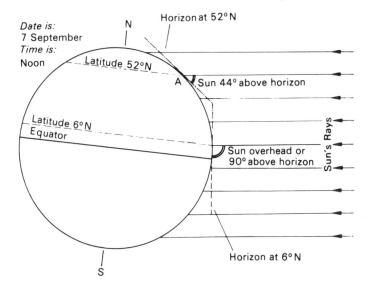

**Fig. 13.** Using a sextant to find latitude. Declination of the sun on a given day = Latitude at which it is overhead at noon. Declination is + if north of equator, — if south. Here declination = +6°. Latitude of A is 90° + declination — angle between sun and horizon: 90 + 6 — 44 = 52°.

equal to about 111 km, so an error of only 1 minute in the time-keeping gives an error of over 27 km in fixing the position. One method of knowing the time depended on measuring the angle between the Moon and certain bright stars, but this was not an accurate method. What was needed was an accurate clock. It was not until the eighteenth century that clocks of the required accuracy were built. Until these became available the mapping of the Earth was a very inaccurate matter. The development of clocks is described in Chapter 7.

## Project 1: Make a Sextant

A simplified version of the instrument used by navigators for centuries. Measures the angle between the sun and the horizon. Use it to find your latitude. This version has 'black' mirrors so you are not dazzled by the sun

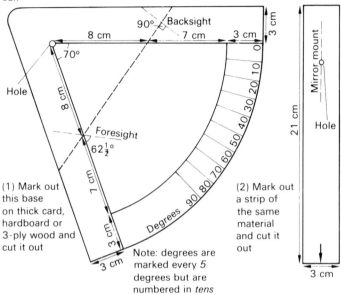

90°  Backsight

8 cm      7 cm      3 cm

3 cm

70°

Hole

8 cm

Foresight

$62\frac{1}{2}°$

7 cm

Backsight

Degrees 90 80 70 60 50 40 30 20 10 0

21 cm

Mirror mount

Hole

3 cm

3 cm

3 cm

(1) Mark out this base on thick card, hardboard or 3-ply wood and cut it out

(2) Mark out a strip of the same material and cut it out

Note: degrees are marked every 5 degrees but are numbered in *tens*

(3) Make the back sight of thin card

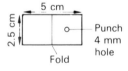

5 cm

2.5 cm

Punch 4 mm hole

Fold

(4) Make the foresight of thin card. Part is blackened with felt-tip pen

Cut out window

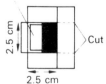

2.5 cm

Cut

2.5 cm

(5) Cut out mirror mount. Part is blackened

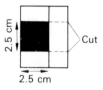

2.5 cm

Cut

2.5 cm

(6) Cut pieces of thin transparent plastic (slide box, cassette box—the hard, glossy 'glassy' kind) to fit black areas (2.5 cm x 2.5 cm, 2.5 cm x 1.25 cm)

(7) Stick on black areas with *plenty* of Bostik 1. Keep glue off front surface

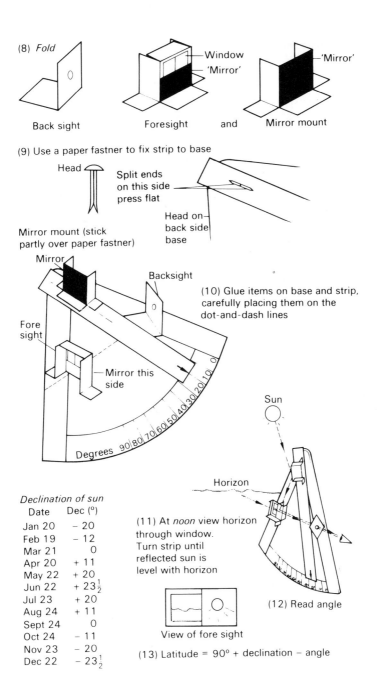

(8) *Fold*

Back sight — Foresight — and — Mirror mount

Window 'Mirror' / 'Mirror' / 'Mirror' / Mirror mount

(9) Use a paper fastner to fix strip to base

Head

Split ends on this side press flat

Head on back side base

Mirror mount (stick partly over paper fastner)

Mirror

Backsight

Fore sight

Mirror this side

(10) Glue items on base and strip, carefully placing them on the dot-and-dash lines

Degrees 90 80 70 60 50 40 30 20 10 0

Sun

Horizon

**Declination of sun**

| Date | Dec (°) |
|------|---------|
| Jan 20 | − 20 |
| Feb 19 | − 12 |
| Mar 21 | 0 |
| Apr 20 | + 11 |
| May 22 | + 20 |
| Jun 22 | + 23$\frac{1}{2}$ |
| Jul 23 | + 20 |
| Aug 24 | + 11 |
| Sept 24 | 0 |
| Oct 24 | − 11 |
| Nov 23 | − 20 |
| Dec 22 | − 23$\frac{1}{2}$ |

(11) At *noon* view horizon through window. Turn strip until reflected sun is level with horizon

(12) Read angle

View of fore sight

(13) Latitude = 90° + declination − angle

*Measuring the mass of the Earth*
Simple geometry was of no help in finding the mass of
the Earth. Before we could measure this we needed some
scientific ideas and some highly precise equipment. The
ideas were those of Newton who, in the 17th century,
united the ideas of Galileo, Kepler and Halley with his
own into one Law of Gravitation. The law deals with the
gravitational attraction between a pair of objects. Each
attracts the other with equal force. For example, if you
hold an apple in your hand you can feel the downward
force (about 1 newton) with which the Earth pulls on the
apple. At the same time the apple attracts the Earth with
an upward force of 1 newton. The size of the force
depends on the masses of the two objects. The force
between Earth and an iron cannon-ball is greater than
that between the Earth and an apple because the cannon-
ball has a greater mass than an apple. Similarly, if you
stand on the surface of the Moon, holding an apple, the
force on the apple is less than 1 newton because the
Moon has a smaller mass than the Earth. The force also
depends on how far apart the two objects are. If we take
our apple a little way out into space the force of the Earth
upon it (and its force on the Earth) is less. The same effect accounts for the behaviour of Richer's clock. The
whole law can be written as an equation:

$$\text{Force} = \frac{\text{Mass of one body} \times \text{Mass of other body}}{(\text{Distance between the bodies})^2} \times G$$

G is a constant number, the Gravitational Constant. One
body might be the Earth, the other body might be the apple we are holding. We can use this equation to find the
mass of the Earth. If we know the *force* on the apple, and
the *mass* of the apple and the *distance* between the apple
and the Earth, *and if we know the value of G*, we can
then calculate the one unknown item in the equation —
the mass of the Earth. It is easy to find the mass of the
apple. Since the size of the Earth has been measured, as
described earlier in this chapter, the distance between its

centre and the centre of the apple is easy to calculate. The only other thing we need to know is the value of G. This was first measured by Henry Cavendish at the end of the 18th century. Almost a hundred years later (1895) an improved version of the same method was used by C. V. Boys. His apparatus measured the gravitational force between two sets of massive bodies. Even though the bodies were massive, the force was only a few thousandths of a newton, but the method is precise and made it possible to calculate a reliable value for G. Since then, G has been measured with even greater precision by Heyl.

Once G had been measured, all the values in the gravitational equation were known except the mass of the Earth. By using all the *known* values, the one *unknown* value can be calculated. The mass of the Earth comes out to almost six million million million million kilograms. No wonder it has such a strong gravitational attraction on all those subjects that are on its surface or near it!

*The age of the Earth*
It was difficult enough to measure the mass of the Earth, but to measure its age with even a small degree of precision has been impossible until recent times.

One of the first persons to attempt to estimate its age by a scientific method was Georges Louis Leclerc, Comte de Buffon, in the 18th century. He believed that long ago a comet passing by the Sun had pulled out masses of molten white-hot material. These cooled down, became solid and then circled the Sun as planets. If the Earth had had this history, it must once have been white hot and had been losing heat ever since. He experimented with spheres of various substances, making them very hot and then seeing how long they took to cool. From his results he estimated how long it should have taken for the Earth to cool to its present temperature. Allowing for the heat that the Earth had received as radiation from the Sun as it cooled, Buffon's estimate of the age of the Earth was 74,832 years. We now know that this figure is far too small, but it was at least a reasonable deduction from carefully made measurements.

Another method of estimating the Earth's age was based on the work of geologists, such as James Hutton and William Smith. They attempted to find out the age of the rocks, for the Earth must be at least as old as its oldest rocks. Rocks are formed when older rocks of hills and mountains are worn away by wind and water and the materials are laid down elsewhere, as sediment. Such deposits are laid down, for example, where a river laden with sediment enters the sea. Observations in areas such as this give an estimate of how long it takes for layers of a given thickness to be built up. If we measure the thickness of layers of rock, we can estimate the ages of the rocks and the fossils found in them. It was calculated that these rocks were several millions of years old. The age of the Earth must be much greater than this for, before such rocks could have been formed, the Earth must have cooled down and the first hills and mountains formed. These measurements showed that the age of the Earth must be reckoned in millions of years.

The key to the age of the Earth was the discovery of radioactivity. Certain elements, such as uranium, exist in forms that are radioactive. The atoms of such elements are not stable. At any instant the nucleus of the atom is liable to split or throw out a particle. We say it *decays*. When it decays, the atom becomes an atom of a different element. When an atom of uranium decays, the atom formed from it is radioactive too. This in turn decays several times turning into other radioactive atoms, and eventually becoming an atom of lead. The lead atoms are not radioactive — they are *stable* and do not decay further. If you were able to watch one particular uranium atom you could not tell when it was about to decay. It might decay immediately, it might decay tomorrow, or next week or maybe not for thousands of millions of years. If you have a lump of uranium or a piece of rock containing uranium, you have millions of uranium atoms. Although the behaviour of *any one* of them is a matter of chance, the behaviour of *the whole* of them follows definite rules. In a given length of time, which we call the half-life, exactly half of the atoms decay. The other half are unchanged and remain as uranium atoms. During the next half-life period, half of those uranium atoms decay, and half are left. Half of a half is a quarter, so now only a quarter of the original uranium atoms remain. In the next half-life, half of that quarter decay, leaving an eighth of the original number. As the amount of uranium decreases, the amount of lead increases.

The length of half-life is fixed for each element. A radioactive form of phosphorus (P-32) has a half-life of 14.3 days. If you are given some P-32, you will find that in about a fortnight half of it has decayed into sulphur. In another fortnight half of the remaining P-32 has decayed into sulphur. In another fortnight only an eighth of the P-32 will be left, and so on. Some elements have a half-life of only a few seconds or minutes; others have half-lives of years. Uranium-238, the commonest kind of uranium found in rocks, has a half-life of 4,510,000,000 years.

One important point about radioactive decay is that it happens at a *constant* rate. There seems to be nothing we can do to give an element a shorter or longer half-life. Half-life is a natural yardstick.

When uranium-238 decays it eventually turns into a form of lead known as lead-206. In young rocks, those only a few hundreds of millions of years old, only a small proportion of the uranium-238 will have decayed into lead-206. In older rocks, that are many hundreds of millions of years old, a much larger proportion of uranium will have turned into lead. The older the rock, the more lead there will be in it, in proportion to the amount of uranium. Since we know the half-life of uranium we can work out how long it has taken for the uranium to decay to lead. This tells us how old the rock is.

Various rocks have been tested in this way. The oldest are found to be about 3,500,000,000 years old. Rocks as old as these have been found to contain remains of some of the earliest living things. This means that the Earth must be even older than this. Additional evidence from other measurements suggests that the Earth is about 5 thousand million years old.

The use of the half-life of radioactive substances as a yardstick for measuring lengthy times has other applications. A half-life of 4,510,000,000 years is rather a large yardstick for measuring the age of the Earth's younger rocks. Instead we measure the decay of potassium-40 into argon-40. This process has a half-life of 1,300,000,000 years. This technique is also useful for dating rocks that contain little or no uranium.

When cosmic rays enter the upper atmosphere they cause some of the nitrogen atoms of the atmosphere to be turned into carbon-14. This form of carbon is radioactive. It combines with oxygen, forming carbon dioxide. This means that the carbon dioxide of the atmosphere is mainly built from carbon-12, with a small and constant proportion of carbon-14. Living plants take up carbon-dioxide and build it into their bodies. The carbon in the body of the plant contains this same small proportion of carbon-14. When the plant dies, there is no further intake of carbon dioxide from the atmosphere. If the remains of the plant are preserved, for example in wood from a tree, the carbon-14 gradually decays radioactively with a half-life of 5570 years. Gradually the proportion of carbon-14 to carbon-12 decreases. For about 70,000 years a measurable amount of carbon-14 remains in the wood. If we test the wood to find out how much carbon-14 and carbon-12 it contains, we can tell how long ago the tree lived. In this way we have been able to measure the age of timber taken from ancient Egyptian tombs. We are also able to measure the age of animal remains such as fossil bones. For example, early fossils of Neanderthal Man, discovered in caves in northern Iraq, have been found to be about 50,000 years old. On the island of Saipan, one of the Mariana Islands of the Pacific, remains of ancient pottery have been found. A shell that was found with the pottery has been dated by the carbon-14 method. It was found to be about 3,500 years old, proving that the peoples living in that area had developed the skills of making pots as early as 1500 B.C. A yardstick by which we can date rocks, and the remains of living organisms is a most useful one for tracing the early history of the Earth and its inhabitants.

**Chapter 4**

# Precision

In Chapter 1 we saw that trade was one of the main reasons for the invention of systems of measurement. Farmers, governors, lawyers, the military men and sailors were also among those who needed to use measurements in their work.

They were not too bothered about the precision of their measuring instruments. If the scales were crudely made and the weights were not exact, the honest trader could easily put an extra fruit on the pan to make the scale swing fully over. The customer would be satisfied and come back another day to buy again. In any case, when there were so many roughly defined measurements in use, what was the point of worrying too much about precision?

Perhaps the people most concerned with precision were the navigators of sailing ships. They needed precise and accurate instruments so that they could find out exactly where they were on the high seas. As we shall see in Chapter 7, their needs led to the invention of highly accurate clocks.

The work of scientists generally depends on being able to make precise and accurate measurements. By inventing better measuring instruments and by thinking of better ways of using them, we have been able to find out more and more about the workings of everything in the Universe.

In the paragraphs above the phrase 'precise and accurate' was used twice. Often we use these words as if they mean the same thing, but each has a quite different meaning. If we say 'This is a precise clock' it does *not* mean the same thing as 'This is an accurate clock'. In this book we often refer to accuracy and precision, so it is important to be sure exactly what these words mean.

*A precise measurement*

To help explain the difference, think about the following example. Suppose you are given a wooden measuring scale, like the one in the drawing. (Fig. 14). It is marked in centimetres. You are asked to use this scale to measure the length of a metal bar to the nearest millimetre. You put the scale on the bar and slide it until the zero end of the scale is level with one end of the bar (Fig. 14). Then you look at the other end of the bar and note which part of the scale is level with the end of the bar. You estimate tenths of a centimetre and record the result: 5.3cm.

If you repeat this measurement several times with the same scale and bar, you will probably find that you do not get the same result every time. The results might be: 5.3cm, 5.5cm, 5.4cm, 5.4cm, 5.7cm, 5.3cm, 5.5cm, 5.6cm, 5.2cm and 5.5cm. Obviously these results can not all be correct. Some or all of them are in error. What are the reasons for these errors? There are several possible causes. One is that you may not have placed the zero of the ruler *exactly* at the end of the bar each time that you made the measurement. Another is that the reading you get at the other end of the bar depends on exactly where you place your eye as you take the reading. This is the effect of *parallax* (Fig. 14).

Both of these errors are caused by *you*, the observer, not using the scale properly. Maybe there are further 'observer errors' due to your making a mistake in reading the scale or in writing down the result. 'Observer errors' of these kinds are *random errors.* The readings

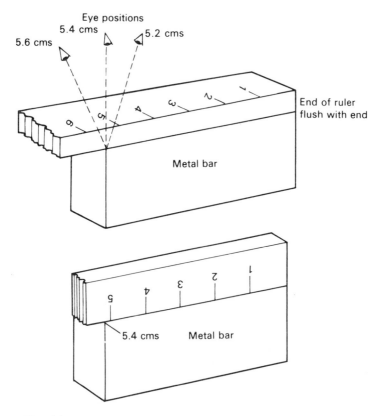

**Fig. 14.**

obtained can be either greater than or less than the true length of the bar. An average of the readings probably gives a good estimate of the length of the bar: 5.4cm. If the marks on the scale are thick or the ruler is old and its markings are partly worn away, you may have great difficulty in making any kind of reading at all. This adds to your random errors. Your measurement is *not precise*, because *random errors are large*.

To make the measurement more *precise* you must try to reduce random errors. Place a block against the ends of the scale and bar to make certain that the zero of the scale is exactly level with one end of the bar. Place the scale on edge (Fig. 14) to eliminate parallax errors. Use a scale marked in millimetres; this scale is more sensitive to length than the centimetre scale. Use a good quality scale with fine clear markings. Then the random errors will be so small that you will get the same reading every time. Your observations will be very *precise*.

A measurement is precise if random errors are small. Random errors may be caused by your measuring methods, or by lack of sensitivity in the instrument you are using. Another cause of random error, and lack of precision is that the length of the object you are measuring is for ever changing. Imagine trying to measure the length of a live earthworm! No matter how careful you are and no matter how good your scale, you can never hope for a precise measurement.

*Accuracy*
In Fig. 15 we see what happens when the scale given to you is precise, but is a badly used one, worn away at the ends. The end of the scale is now at the 1mm mark. As well as random errors such as we have already described, there will now be a *systematic error*. A systematic error always operates in the same direction, either making *all* the readings too big, or making them *all* too small. In this example, systematic error makes all the readings 1mm too big. This error adds on to the various random errors so that your readings are scattered around a value that is 1mm greater than the true value. Even if they are very precise, they are *not accurate*.

If the scale is warped, we get another kind of systematic error. Even if the zero of the scale is accurate, all readings will be too large (Fig. 15). The amount of error is in proportion to the length being measured. With cheap scales, even if they are straight, you may find that *all* the marks are too far apart or too close together, simply because they are badly made.

A third kind of systematic error could be caused by you. Using a centimetre scale and trying to estimate the

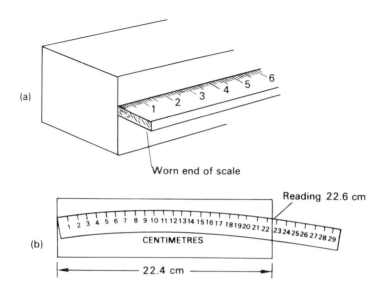

**Fig. 15.** Inaccurate scale.

millimetres, you might tend to over-estimate the distances in millimetres. Or you might tend to under-estimate them. In either case you would be making a systematic error, leading to loss of accuracy.

To make an *accurate* measurement you must reduce all *systematic* errors. Use a scale that has its zero mark intact. Make sure that the scale is not warped, and that it is properly marked. Try not to cause systematic errors by your way of using the scale.

To sum up, a measurement is *precise* if *random* errors are small. It is *accurate* if *systematic* errors are small. A digital watch with a display that reads to the nearest 1/100 of a second is precise. If it has been set exactly to the Greenwich time signal and it gains or loses only a few seconds a month, it is precise *and* accurate. If its display reads only in hours and minutes, it may be very accurate, but it would not be called precise. If you set it wrongly to begin with, it might be precise, but is not accurate.

*Standards*

For a long time we have based our units of length on a *standard*. An early standard was King Edgar's yardstick. The *Imperial Standard Yard* was made in 1856. It was a bar of a special type of bronze with a gold plug sunk into the bar near each end. On each plug was a fine mark. The distance between the two marks when the bar was at temperature of 62°F was defined by law as one yard. This standard was designed to be *precise*; the marks were very fine and it was used at a fixed temperature to eliminate errors due to expansion. It was also *accurate*; it was made of metal that would not gradually change in length; the marks were on gold; which does not corrode; it was always supported in a certain way to reduce errors caused by bending under its own weight.

The *International Prototype Metre* was similarly precise and accurate. It was made from a special platinum-iridium alloy so the fine marks could be made directly on the bar itself. The bar had a special X-shaped cross-section to make it very rigid yet light in weight. It was always used at 0°C.

Naturally, these standards were irreplaceable and needed to be given special care. They were kept safely and used as seldom as possible. The Imperial Standard Yard was kept at the Board of Trade in London. Five copies were made and kept at the Royal Mint, the Royal Society, the Royal Greenwich Observatory, the Board of Trade and at the Palace of Westminster. The first four copies were compared with each other every 10 years by the National Physical Laboratory, and were each compared with the Imperial Standard every 20 years. In this way we had a precise and accurate yardstick to use for marking *accurate* copies, or *sub-standards*. These would be copied again and used when marking out the rulers and scales we use in the laboratory or in everyday life.

In recent years it has been realized that a standard may change over a very long period of time. There is also the risk that it may be lost or destroyed. Scientists have felt happier with the idea of using natural standards of length instead of one defined by the distance between two marks on one particular bar of metal. Instruments are so precise nowadays that the idea of using our hand or foot as a natural standard is no longer acceptable. Instead, we use the wavelength of light. In the International System the metre is defined as the length of 1 650 763.73 wavelengths of red light of a particular band of the spectrum being emitted by atoms of krypton-86. Since the wavelength of light is so small, the metre can be defined with very high precision. A special instrument is used to compare sub-standard bars with the wavelength of krypton light and to measure the length accurately and precisely. Now the standard is not a bar of alloy but a beam of light. The metre is now defined by reference to light waves, and the yard is now defined as 0.9144 metres.

*Yardsticks of mass*
Just as we had a specially-made standard length, we have specially-made standard masses. The Imperial Standard Pound, was a particular cylinder of platinum kept and copied in the same way as the standard yard. The International Prototype Kilogram is a cylinder of platinum-iridium alloy kept at the Bureau International de Poids et Mesures, at Sévres, France. The SI unit of mass is still defined as the mass of this prototype kilogram. The Standard Pound is no longer used. Instead, the pound is defined as 0.45359237 kilogram.

*Yardsticks for time*
Until the early years of this century it was believed that the Earth rotated on its axis at a constant rate. It was thought to be a precise and accurate standard of time. This defined the *day*, which was divided into 86400 seconds. There are several different definitions of day, depending on exactly what is meant by 'one rotation', but for all of these the Earth was considered to be the standard. In 1925 a new type of pendulum clock was in-

stalled at the Royal Greenwich Observatory. It was then found that the period of the Earth's rotation varied by amounts up to 0.003 seconds per day. This was caused by *nutation* which is a kind of slight nodding motion of the Earth's axis. This had been discovered by James Bradley in the 18th century. Now it was found to be making the Earth no use as a precise timekeeper. When clocks of even greater precision had been invented (atomic clocks), it was found that the Earth's rotation also has a systematic error. It is becoming slower, and the length of the Earth's day is decreasing by about 1 second each year. The Earth was neither precise nor accurate.

The atomic clocks are the most precise clocks we have. They work by detecting the vibrations of caesium-133 atoms when they are made to produce high-frequency radio waves. The vibrations can be counted electronically. The vibrations that produce radio waves of one particular wavelength are now used as our standard of time. The second is defined as the time required for 9 192 631 770 such vibrations of the caesium atom.

It is interesting to note that we are defining more and more of our units by comparing them with *natural* yardsticks. Length and time are now defined by the radiation produced from atoms. Mass is still defined by the man-made prototype kilogram but maybe one day a natural yardstick will be used to define mass too. The advantage of a natural yardstick is that it does not depend upon just one particular object that can be lost or destroyed, or may change with time. Given some caesium, the necessary materials, the skill, and the knowledge, a scientist can build a caesium clock and use it as a precise time reference. This can be built at any convenient place on Earth, or even in some other part of the Universe. It can be built now or at any time in the future. To define our units of time and length, all we need are a few sheets of paper giving the instructions for building and using the equipment. The units depend only on the atoms themselves. We assume that the properties of atoms are the same wherever they may be and will remain the same for ever. If this is so, the yardstick is a universal and permanent one.

*Improving precision*

An engineer's scale is a precise instrument for measuring length. The scale is engraved with very fine lines, 0.5mm apart. By using a magnifying glass to take the readings we can measure a length to the nearest 0.1mm. The magnifying glass helps us see the small distances between the end of the object and the nearest marks on the rule. Magnification or *amplification* is a good way of improving precision. Another example is the mirror galvanometer (Fig. 16), used to measure very small electric currents. The coil turns through a small angle when a current passes through it. A mirror is fixed close to the

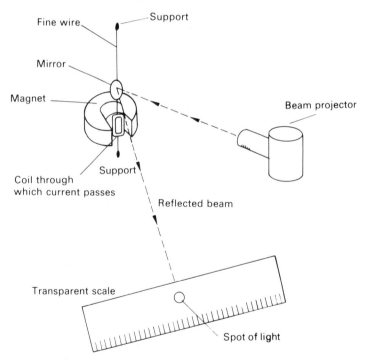

**Fig. 16.** How a mirror galvonometer works.

coil and turns with it. As it turns the reflected spot of light moves across the screen. Because of the laws of reflection, the angle turned by the reflected beam is always exactly *twice* that turned by the coil. This gives us twofold magnification. In addition, the scale of the instrument is placed a long way from the mirror. A small change in the angle of the beam makes the spot move a long way across the scale. This provides even more magnification. For a typical mirror galvanometer, an increase or decrease in current of 1$\mu$A (1 microampere = 1 millionth of an ampere) causes the spot to move 500mm along the scale. Does this mean we can measure currents with a precision of one five hundredth of a microampere? Does this mean that we can increase precision even more by taking the scale even further from the mirror? If you watch the spot, you will soon see the reason why it is *not* possible to read the scale to 0.02$\mu$A or to further increase the precision. The spot is not absolutely still but is jumping about. When we magnify the motion due to electric currents we also magnify motion due to other causes. Draughts and winds would blow the mirror around badly, but these are almost eliminated by the case around the instrument. Even so, there are small vibrations from the surroundings that make the mirror quiver. This produces very *small* random movements of the mirror that give *large* random movements of the spot. It is not possible to be certain of the exact position of the spot on the scale. We have reached the limit of precision of the galvanometer.

This example illustrates one of the most serious problems in measuring. This is the problem of *noise*. We use this word not in the sense of actual noise, as a kind of sound, but in the sense of *unwanted disturbance*. With the mirror galvanometer, vibrations are caused by people walking around in the room or by traffic in the street, or by strong winds shaking the buildings. These vibrations are noise because they are unwanted disturbances of the mirror. When we magnify a quantity to increase the precision with which it can be measured, we often magnify the noise too. This is what happens with the galvanometer. So we try to reduce noise as much as possible. We put it in a case to cut out draughts. We place the galvanometer on a very firm and massive workbench, to reduce vibrations. But we are always likely to be left with a certain amount of noise in any system and this limits the precision of our measurements. A good example of a way of overcoming the noise problem is illustrated by the speckle interferometer, described on p.76.

## Project 2: Make a Steelyard

A kind of balance first used by the Romans. Since it needs only temporary use of only one standard weight it had an advantage in days when accurate standard weights were scarce.

(1) Cut out this shape from stiff card, to make the pan

(2) Cut out a cardboard 'handle'

Hole

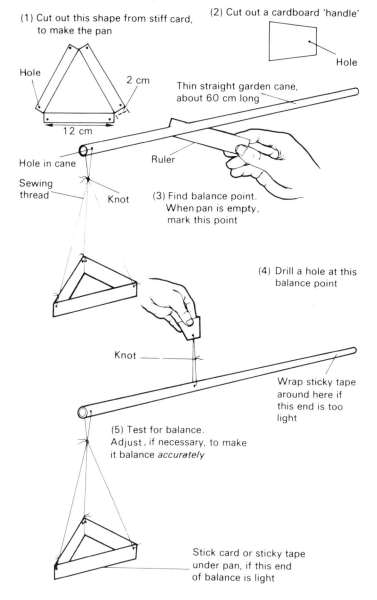

Hole

2 cm

Thin straight garden cane, about 60 cm long

Hole

12 cm

Hole in cane

Ruler

Sewing thread

Knot

(3) Find balance point. When pan is empty, mark this point

(4) Drill a hole at this balance point

Knot

Wrap sticky tape around here if this end is too light

(5) Test for balance. Adjust, if necessary, to make it balance *accurately*

Stick card or sticky tape under pan, if this end of balance is light

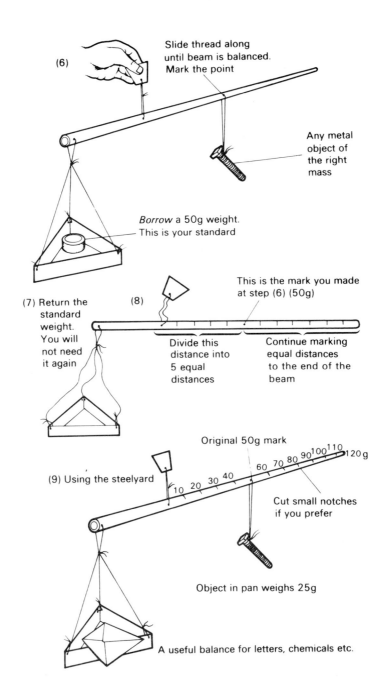

(6) Slide thread along until beam is balanced. Mark the point

Any metal object of the right mass

*Borrow* a 50g weight. This is your standard

(7) Return the standard weight. You will not need it again

(8) This is the mark you made at step (6) (50g)

Divide this distance into 5 equal distances

Continue marking equal distances to the end of the beam

(9) Using the steelyard

Original 50g mark

10  20  30  40  60  70  80  90  100  110  120 g

Cut small notches if you prefer

Object in pan weighs 25g

A useful balance for letters, chemicals etc.

**Chapter 5**

# Measuring the Universe

The first person to make a scientific measurement of distances beyond the Earth was Aristarchus of Samos in the 3rd century B.C. When the Moon is exactly half-illuminated, as seen from the Earth, the Sun-Moon-Earth angle is exactly 90° (Fig. 17). He tried to measure the Moon-Earth-Sun angle and obtained the value 87°. Calculation then gives the result that the Sun is 19 times further from the Earth than the Moon is. The disadvantage of his method is that when angles close to 90° are to be measured, a small error in measurement leads to a large error in the final result. In this instance, the angle was actually 89.85°, and the Sun is actually 389 times more distant than the Moon, on average. He had, however, established that the Moon is much closer to us

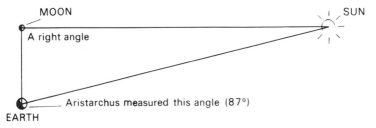

**Fig. 18.** Aristarchus measures the relative distances of Sun and Moon.

than the Sun. Since both Moon and Sun appear to be almost the same size when viewed from Earth, Aristarchus reasoned that the Sun must be much bigger than the Moon.

He tried to work out how much bigger it was by making observations during an eclipse of the Moon. He estimated that the diameter of the shadow of the Earth was about 2⅔ times the diameter of the Moon. If this is so, and we assume that the Sun's rays are more or less parallel, the diameter of the Earth is about 2⅔ times that of the Earth. This agrees fairly well with the actual figure, which is 3.6 times. Aristarchus then concluded that since the Sun is 19 times further away, its diameter is about 7 times that of the Earth. His under-estimate of the distance of the Sun led him to underestimate its diameter, which really is about 109 times that of Earth.

Another and perhaps the greatest of ancient astronomers was the Greek, Hipparchus of Nicaea, who lived in the 2nd century B.C. Among his achievements was the measuring of the length of the year with an accuracy of 6 minutes. His main contribution to astronomy was to make a catalogue of the main stars. He measured their positions on the celestial sphere, as celestial latitude and longitude. The celestial sphere is the imaginary sphere in which all stars appear to lie, as viewed from Earth. Its north and south poles are directly in line with the Earth's poles and it appears to rotate around these poles once in 24 hours.

The celestial equator is the projection of the Earth's equator on to the celestial sphere. Such a system of locating star positions is still used today. Hipparchus, using a crude sighting instrument, was able to measure star positions to the nearest 20 minutes of arc (⅓ of a degree). Even with this low precision he was able to note that the positions of stars were not the same as those recorded by earlier astronomers. He concluded that the Earth's axis was slowly altering in direction just like that of a spinning top. This effect he called *precession*. It was Hipparchus who invented the system of magnitudes, for describing the brightness of stars (p.18). Later, the star lists of Hipparchus became the basis of the much larger catalogue prepared by the Alexandrian astronomer, Ptolemy (2nd century A.D.). In this list, generally known as the *Almagest*, he listed star positions with a precision of 10 minutes of arc though, as he used the same kind of instrument as Hipparchus, his measurements were probably accurate only to 20 minutes. It was not until the work of Tycho Brahe that there was any significant improvement in the precision of measurements of star position. Tycho Brahe was a Dane living in the 16th century. He realised that scientific discoveries depended on precise and accurate observations. He used a quadrant made from steel. It had a precision of 4 minutes of arc. He devised techniques to keep errors to a minimum. As a result of this, his measurements were as accurate as is possible by use of the naked eye alone. The next great improvement in precision came with the invention of the telescope. The English clergyman John Flamsteed, appointed in 1675 as the first Astronomer Royal at the newly founded Royal Observatory at Greenwich, used a telescopic sextant to measure the altitudes of the stars as they crossed the meridian. With this instrument he was able to achieve a precision of 10 *seconds* of arc. Now that such precision was attainable, further discoveries and measurements came quickly.

*Measuring by parallax*
We have mentioned that parallax can be a source of random error when a scale is being used to measure lengths (p.4). Parallax has its uses in measuring the distance of the Sun and the nearer stars. If you hold an object such as a pen or pencil or even a finger upright in front of your eyes at about six inches, the effect of parallax can be seen when you open and close each eye alternately. The object will appear to move from side to side relative to the background scene. We learn to interpret the different views seen with each eye and use the information

to judge the distance between the nearby object (the pencil) and our eyes. The Italian astronomer, Cassini, who later took French nationality, measured the parallax of Mars. When we measure the parallax of a planet we assume that the stars are so far away that they can be thought of as a fixed background. A planet viewed from two different positions is like the pencil mentioned above. In 1672, Cassini viewed Mars from two different positions on Earth. He measured the angles between Mars and some bright stars that were close to it in the sky, as seen from Paris. Similar measurements were made at the same time from Cayenne, by Jean Ficher. Paris and Cayenne are nearly 1000 km apart, so Mars appears in a different position against the background of stars when viewed from these two places. From these measurements the distance between Earth and Mars was calculated. Kepler had already published his laws, one of which states the relationship between the size of a planet's orbit and the period in which it makes one orbit. The periods were known, so Kepler's Law had already established the *relative* sizes of the orbits of the planets. Now that Cassini had actually measured one of the distances, the *scale* of the Solar System was known as well. The *actual* sizes of the orbits of all known planets could then be calculated. In particular, the length of the Astronomical Unit, the average distance between Earth and Sun, could be calculated. From Cassini's observation this was found to be 149,500,000 km. This figure agrees well with the modern value 149,597,910 km, which was obtained by timing pulses of radar waves sent from Earth and reflected back from Venus.

As instruments became more precise it became possible to measure the parallax of the nearest stars, and so to find their distances. An instrument called a heliometer was the first one used for this purpose. The first heliometer had been built in 1754 by John Dolland, lens maker. It was designed to measure the diameter of the Sun's disc, as seen from Earth. It was similar to a telescope, except that the lens was split into two D-shaped halves. These halves could be moved so that each produced an image of a slightly different part of the sky. By moving the lenses the images of two stars could be brought into line. Then the amount by which the lenses had been moved indicated the angular distance between these stars. With such a precise way of measuring the angles between stars, the German astronomer, Friedrich Bessel, began to measure the positions of certain stars. It has already been observed that some of them seemed to change their position regularly during the year. This is due to parallax as the Earth circulates in its orbit (Fig. 18). Using this method Bessel calculated the distance of the star known as 61 Cygni. His measurements were precise, for the angular change was less than one third of a second arc. Calculation gave the distance of 61 Cygni as 10.3 light-years. A light-year is the distance travelled by light in 1 year (see Chapter 6), which is approximately 9.5 million million kilometres. Since the time when Bessel made his observation, the invention of photography has made it possible to measure parallax with even greater precision. Photographs of the skies

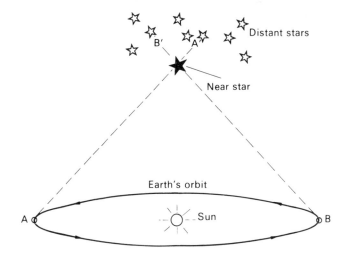

taken from different positions in the Earth's orbit can be compared and angular changes measured. The median value for the distance of 61 Cygni is now reckoned as 11.2 light years, a figure close to Bessel's. The distances of thousands of the nearer stars have been measured by the parallax method.

The measurement of the distance of 61 Cygni was the first measurement of a distance outside the Solar System. It gave us our first indication of the immense size of the Universe. Even so, the stars so measured are very much closer to us than most others. New techniques were needed to measure these even greater distances. One of these has some points in common with the parallax method. This is the moving cluster method. Imagine a lorry moving at constant speed along a straight road at night. It has a cluster of lights at the rear. It is a dark night, so you can see the lights but not the lorry or the road. You can not see the lorry so you do not know how big it is. You can not tell how far away it is by seeing how big it looks. It might be a small van near to you or a juggernaut lorry a long way away. If you watch its lights carefully, you can see them converging on a point in the distance. If the lorry is near, the cluster of lights moves rapidly past your view. If the lorry is further away, the group as a whole appears to move slowly. At the same time as this change occurs, the lights appear to move closer and closer together as they all converge on that distant point at the end of the road. Just by watching how fast the *whole cluster* of lights appears to move, and how rapidly the *individual lights* appear to converge, we can estimate the distance of the lorry. We can estimate the distances of moving clusters of stars by a similar method. The cluster *as a whole* appears to move across the background of more distant stars and the *individual stars* appear to move closer to each other. By making measurements we can estimate the distance of the cluster. This method showed that the distance of our nearest cluster, the Hyades, is about 148 light-years.

*Measuring by magnitude*

If someone were to put several identical lighted candles in different parts of a large field at night, you would easily be able to tell which were nearest to you and which were further away. The nearer candles would appear brighter than the distant ones. When we look at the stars, some appear brighter than others. We might assume that the brighter ones are the nearer ones, but often we would be wrong. Unlike the candles, stars are not all alike; they do not all give out the same amount of light. A star that is really very bright, but is far away may appear from Earth as a very faint star. Yet a weakly shining star that is closer to us may appear relatively bright. This is the difficulty that arises when we try to use magnitude to estimate distances. When we look at a star, how can we tell its true brightness, or *absolute magnitude*? Fortunately there is one type of star that allows us to estimate its absolute magnitude fairly easily. Some kinds of stars vary in brightness from time to time. We call them *variable stars*. They go through periodic increases and decreases in brightness. When they are at their brightest they may be as many as 2 magnitudes brighter than when they are at their dimmest. Most vary in brightness at a regular rate. Some go through their cycle in a day or less; others take several weeks or even years. One particular type of variable star is the cepheid variable. As they change in brightness, cepheids change in size and temperature too. It is possible to recognize cepheid variables in our own galaxy and also in other galaxies. At the beginning of the 20th century the American astronomer Henrietta Levitt, sudied the variable stars in a galaxy not far from our own, the Lesser Magellanic Cloud. Several of these were variables of the cepheid type. Since they are all in the same galaxy, they are all at approximately *the same distance* from us. Yet they are not all of equal brightness. Further observations showed an interesting fact: the longer the period of the brightness cycle of a given star, the greater is its average brightness. Cepheid variables having the *same period* have the *same average brightness.* For example in the Lesser Magellanic Cloud, the cepheids with a period of about 70 days are all of magnitude 12. Those with a period of about 20 days are all of magnitude 14. After this discovery of Levitt's, another American astronomer, Harlow Shapley, measured the distances of two cepheids in our own galaxy. He found that these too obey the brightness rule. Since we know the distances of the cepheids in our galaxy we can work out their true brightnesses. If we then look in the Lesser Magellanic Cloud for cepheids having the *same period*, we know that they have the same true brightness as the cepheids in our own galaxy. Because of their greater distance, the cepheids in the Lesser Magellanic Cloud all look fainter than the corresponding cepheids in our galaxy. They look about 36 thousand times fainter, yet we know they actually have the same brightness. The loss of brightness tells us how far away they are — a distance of 150,000 light years.

The cepheid variables are all large bright stars (over 1000 times brighter than the Sun) that can be seen at great distances. The way their brightness varies makes them easy to recognise, even in distant galaxies. We can look at a distant galaxy, measure the period of a cepheid and its apparent magnitude and then, from these observations, we can calculate the distance of that galaxy. For measuring distances to distant galaxies, the cepheid variables are one of our most important yardsticks.

Another similar yardstick is the globular cluster. This is a spherical group of up to a million stars, all closely clustered together. Since they contain so many stars, globular clusters are very bright and can be seen at enormous distances. David Hames, a Cambridge astronomer, has examined the clusters in our own galaxy and other galaxies. It is found that there is always the same 'pattern of brightness' in every galaxy. By 'pattern of brightness' we mean that a certain percentage of the clusters is always very bright, another percentage is less bright, and another percentage is least bright. The distances of clusters in our own galaxy are already known. If we then look at a distant galaxy and see how bright the different kinds of globular clusters in it appear to be, we can estimate the distance of that galaxy. In this way distances of millions of light-years can be measured.

*Light-Years*
In this chapter we have been dealing with immense distances. To say that the distance from the Sun to the nearest star is 40,681,010,000,000,000,000 km is meaningless when so many figures are required. Even if we shorten the expression to $4.068 \times 10^{19}$ km we are still left wondering at the largeness of the number. To measure astronomical distances we need longer yardsticks. One larger yardstick is the astronomical unit (p.54) another is the light-year (p.55). In saying that the distance from the Sun to the nearest star (in the Alpha Centauri system) is 4.3 light-years, we understand that it takes 4.3 years for light from this star to reach us. Light travels at a tremendous speed. According to Einstein's Theory of Relativity, nothing travels faster. Before going on to consider other measurements of the Universe we will look at the ways in which the speed of light has been measured. This is the subject of the next chapter.

**Chapter 6**

# Measuring the Speed of Light

As far as we can tell with our senses, light travels very fast. For example, it travels much faster than sound. We see a batsman strike the cricket-ball but the sound takes a noticeable time to reach us. We see a flash of lightning but do not hear the thunder until several seconds afterwards. Light travels so fast that many of the early thinkers, including Aristotle, believed that it travelled instantaneously, that it had *infinite* speed.

The first suggestion for a method of measuring the speed of light was made by Galileo. Two men with lanterns are to stand on hilltops, a few kilometres apart. The lanterns have shutters so that the men can send a flash of light from one hilltop to the other. If we could measure the time between the opening of the shutter of lantern A and the first sight of the response from lantern B, the speed of light could be measured. The experiment was tried on several occasions but always the time was too short to measure. A time that must be measured in microseconds was beyond the grasp of the instruments of the 17th century.

Later in the 17th century the first successful measurement was made. The Danish astronomer Olaus Roemer had been observing the eclipses of the satellites of Jupiter as they passed into the shadow of the planet. It would be expected that these eclipses would occur at regular intervals and that the times of the eclipses could be predicted. For example, the satellite Io orbits Jupiter once in every 42.46 hours so we would expect eclipses to occur every 42.46 hours. It had already been noticed by other astronomers that eclipses do *not* occur regularly. Their times vary during the year, sometimes being early and sometimes late. Roemer noticed that the eclipses were earliest when Earth was at that part of its orbit nearest to Jupiter. About six months later, when Earth was farthest from Jupiter, the eclipses were at their latest. The delay between earliest and latest times was about 22 minutes. Roemer suggested that the delay was due to the extra time taken by light to travel the extra distance across the Earth's orbit. The diameter of the Earth's orbit was then thought to be 290,000,000 km. Twenty-two minutes is 1320 seconds, so the speed of light was calculated as 220,000 km per second. Although this value is too low because of errors in measuring the diameter of Earth's orbit and the time delay, it is a remarkably good result and within range of our present value of 300,000 kilometres per second. Roemer had demonstrated that the speed of light was not infinite. Its speed, though extremely great, was at least measurable.

Another astronomical measurement of the speed of light was attempted by James Bradley in 1728. He had been observing the stars to measure their distances by the methods of parallax (p.55). He noted that when a certain star ($\alpha$ Draconis) was overhead, it showed a displacement from its expected position that could not be due to parallax. This effect was an unexpected one that had not been noted before. It demonstrates how improvement in the precision of measuring instruments can lead to new unexpected discoveries. Bradley explained this effect, the aberration of starlight, as the result of the Earth's motion in its orbit around the Sun (Fig. 19). The Earth's speed in orbit is 30 kilometres per second, and this is large enough to produce a measurable effect. If you are in a stationary car in a rainstorm and there is no wind,

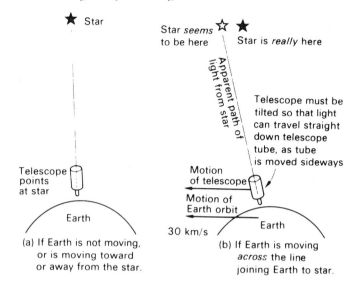

★ Star

Star *seems* ☆  ★
to be here      Star is *really* here

Apparent path of light from star

Telescope must be tilted so that light can travel straight down telescope tube, as tube is moved sideways

Telescope points at star

Motion of telescope

Motion of Earth orbit

Earth

30 km/s

Earth

(a) If Earth is not moving, or is moving toward or away from the star.

(b) If Earth is moving *across* the line joining Earth to star.

**Fig. 19.** Bradley's discovery.

the raindrops are seen to be falling straight down. If the car is moving, the drops appear *to you* to be coming towards you at an angle to the vertical. In the same way, we need to turn our telescope slightly ahead of the star to 'catch' the light coming from it. The angle depends on the speed of the Earth and the speed of light. Bradley already had an estimate of the Earth's speed. He measured the angle and then was able to calculate the speed of light. His result was 295,000 kilometres per second, which is very close to our present value. Later, it was Bradley who, in his appointment as third Astronomer Royal at Greenwich, discovered the nutation of the Earth's axis (see p.45).

The first attempt to measure the speed of light over a measured distance on Earth was made by the French physicist, Fizeau, in 1849. His apparatus used a rotating wheel to send pulses of light to a distant mirror (Fig. 20). If the wheel was still or rotating slowly, a pulse of light could travel to the distant mirror and back, passing

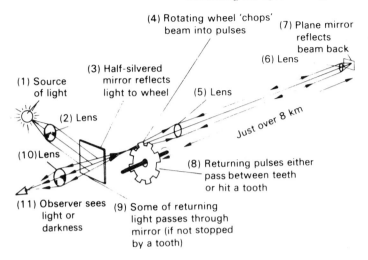

**Fig. 20.** Fizeau's method for measuring the speed of light.

between the same pair of teeth to the eye of the observer. At a certain increased rate of rotation, a returning pulse of light would find that the wheel had rotated enough to put a tooth in its way. At this rate of rotation the observer could not see the returning light pulses. The rate of rotation could be accurately measured so it was easy to calculate how long a pulse took to make the journey *out* between two teeth and to hit a tooth at the end of the return journey. From these measurements, Fizeau calculated the speed of light as 313,000 kilometres per second. This result, and that of another French physicist, Jean Foucault, who used a similar apparatus based on a rotating mirror, were more precise than the earlier astronomical measurements. They could measure distances and rates of rotation precisely and did not need to rely on the rather inaccurate estimates of astronomical distances.

In the early twentieth century a precise measurement was achieved by the American scientist, Albert Michelson. His apparatus worked on the same principle as that of Fizeau (Fig. 21). The distance of the reflector

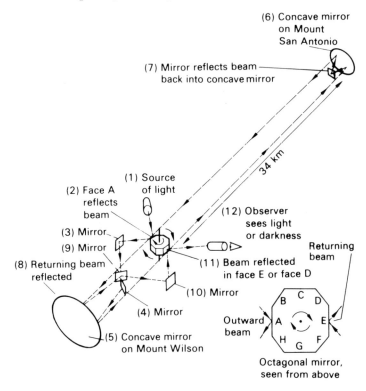

**Fig. 21.** Michelson's method for measuring the speed of light.

was much greater and was measured with a precision of 1 part in ten million. Instead of a toothed wheel, he used a rotating octagonal mirror. With the mirror rotating slowly, a pulse of light leaving face A would be received back on face E. When the mirror was rotated faster, face E had turned out of position before the pulse returned, so the beam was deflected and was not seen by the observer. When the rate of rotation was 528 revolutions per second, a pulse of light reflected from A would be caught by face *D* on its return and reflected exactly into the eyepiece. The time for the return journey was then known to be $\frac{1}{8 \times 528}$ seconds. From several such experiments, the speed of light was calculated as 299,798 kilometres per second.

Michelson also attempted to measure the speed of light in a vacuum, though he died before the measurements could be made. A special tube 1 mile long was used and the mirror was 32-sided. To allow a more precise measurement, mirrors were used to reflect the beam of light along the tube several times. The experiment was completed by his colleagues after Michelson's death. The result obtained was 299,774 kilometres per second.

In recent years different methods have been used to measure the speed of light with increasing precision. One method has been to use the Geodimeter and tellurometer (p.26) over measured distances. If the distance and the number of wavelengths in that distance is known, we can calculate the actual wavelength. Then if we know the frequency, we can calculate the speed. The Geodimeter uses visible light and the tellurometer uses radio waves, but both types of radiation have the same speed. Another method is best illustrated by describing what happens with sound waves. A tuning fork is held at the end of an open pipe (Fig. 22) and the length of the pipe is gradually adjusted until the note from the fork is suddenly heard very loudly. At this point the tuning fork is making the air in the pipe vibrate strongly. We say it is making it *resonate*. When this happens the sound waves from the fork are passing down the tube and being reflected back in such a way that a pattern of *standing waves* is found in the tube. If we measure the tube and know how many standing waves there are, we can find the wavelength and then calculate the speed of sound. It is also possible to generate standing *radio* waves in an enclosed cavity, if short wave radio waves, such as microwaves (radar) are used. Again, if we know the length of the cavity and the number of waves, we can calculate the speed of the radio waves. As mentioned above, the speed of all types of electromagnetic radiation (radio, microwaves, ultraviolet, visible light, infra-red, X-rays, gamma rays) is the

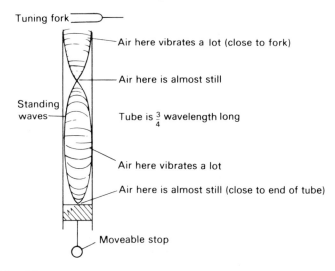

Fig. 22.

same, so the results of this experiment tell us the speed of light too. As a result of such experiments, the speed of light is now known to be 299,792.458 kilometres per second. Over the years, since Roemer's first observations, our value for the speed of light has increased both in accuracy and precision.

## Project 3: Make an Astrolabe

The instrument most often used by ancient astronomers. It was used to measure the angle of elevation of a star or the moon or sun, until the sextant was invented.

(1) Cut out base, about 8 cm diameter. Mark scale in tens of degrees.

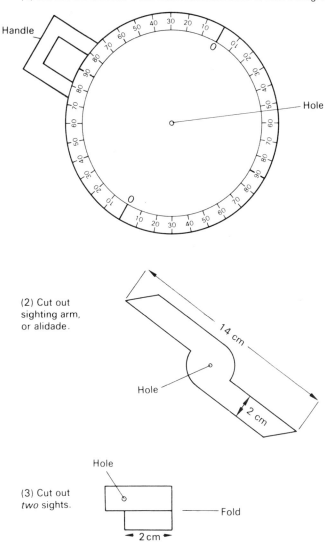

Handle

Hole

(2) Cut out sighting arm, or alidade.

14 cm

2 cm

Hole

(3) Cut out *two* sights.

Hole

Fold

2 cm

(4) Glue
sights to
alidade.
Fasten alidade to base.

Paper fastner

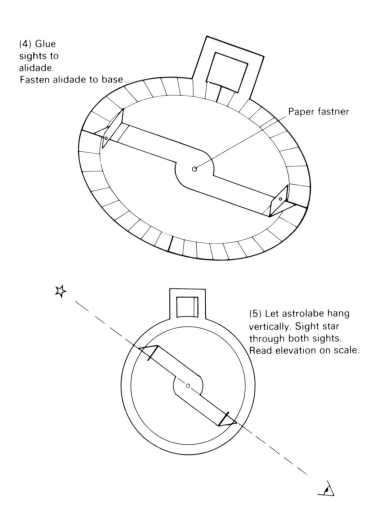

(5) Let astrolabe hang
vertically. Sight star
through both sights.
Read elevation on scale.

(6) The pole star is *almost* directly
above the north pole. If you measure
its elevation, this is almost equal to
your latitude.

Ancient Arabian astrolabes
were marked to show altitudes
of well-known stars at
different times. Their astrolabes
could be used to tell the time at
night.

**Chapter 7**

# The Expanding Universe

The Austrian physicist Christian Doppler gave his name to an effect that we all frequently experience in everyday life. When a fire-engine or ambulance rushes along the street, sounding its syren, you notice a distinct drop in the pitch of the sound as the vehicle passes you. As it approaches the syren sounds are higher in pitch. As it passes away from you the pitch of the sound abruptly becomes lower. We call this the Doppler effect. All kinds of wave motion show the Doppler effect. We can observe it with all the electromagnetic waves, including light. In everyday life we do not notice this effect with light because the moving objects we can see are all moving far too slowly in comparison with the speed of light. A change in the frequency (or wavelength) of sound leads to an alteration of *pitch*. A change in the frequency (or wavelength) of light leads to an alteration of *colour*. When an object approaches us, the Doppler effect causes the frequency of the waves reaching us to be increased (a *de*crease of wavelength): the colours of such an object are shifted toward the violet end of the spectrum. When an object is receding from us, the light reaching us has lower frequency (longer wavelength) and its colours become reddened. This effect is usually known as the *red shift*. Although we are not able to detect a red shift as a supersonic aeroplane disappears toward the horizon, we can certainly detect it when we look at distant stars and galaxies. If we can measure how much reddening occurs, we can calculate the speed of the object as it moves away from us. The only problem is that, before we can tell how reddened its colours are, we need to know what colours it would have if it was *not* moving away from us. If a stationary star was emitting light of all visible wavelengths and ultra-violet as well, it would look white to us. If it started to move away from us, the frequency of all its radiations would be reduced: all its radiations would be shifted toward the red end of the spectrum. Light that appears green as it leaves the star, might appear yellow when it reaches us. Light that appears blue as it leaves the star might appear green when it reaches us. All the colours shift in this way. Red light shifts towards infra-red and becomes invisible to the eye. Ultra-violet shifts toward violet and becomes visible as violet. Although there has been a shift, there is no overall effect as seen by the eye. The star still looks white.

The reason we can detect the shift is that certain wavelengths of light are *marked* as they leave the star. The intensely hot interior of the star is emitting light and other electromagnetic radiation of a wide range of wavelengths. As this light passes through the cooler outer layers of the star, light of certain wavelengths is absorbed by atoms in that layer (Fig. 23). When we look at the spectrum of that light there are dark bands across it at these wavelengths, for no light of those wavelengths is leaving the star. These lines, the *Fraunhofer lines*, are named after the German physicist, Joseph Fraunhofer, who discovered them in the early 19th century. Atoms of different elements absorb light of a number of wavelengths to form distinctive patterns of Fraunhofer lines. Experiments in the laboratory tell us which patterns correspond to which elements. We use this information to find out which elements are present in the outer layers of the Sun and other stars. We look at the spectrum of light coming from the star, note which patterns of lines are present and if the pattern shows up strongly

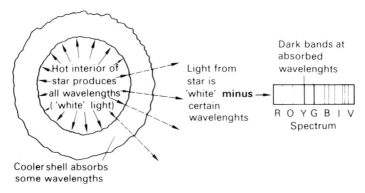

Fig. 23. Production of Fraunhofer lines in a star's spectrum.

or weakly. We can tell not only which elements are present, but the relative amounts of each.

The other use of the Fraunhofer lines is that they act as markers for particular wavelengths. If we examine the spectrum of a star that is moving away from us, the *whole pattern* of lines is shifted toward the red end of the spectrum. Each line is present and is at the proper distance from its neighbours, but each line is nearer the red end of the spectrum than if the star were not moving away from us. This is how we can measure the speed of the star as it recedes from us.

In the early 20th century, Edwin Hubble, an American astronomer, studied the light reaching us from distant galaxies. He observed their red shifts and calculated the speeds at which they were receding from us. He discovered the interesting fact that the more distant the galaxy, the greater the red shift. Measurement showed that there is a definite rule connecting the distance of the galaxy and its speed of recession. This rule became known as *Hubble's Law*. The speed of recession of a galaxy is about 16 km/s for every million light years of distance. For example, a galaxy 100 million light-years away would be leaving us with a speed of 1600 km/s. The most distant object measured so far is the quasar OQ 172 observed by Margaret Burbridge, the British astronomer and a director of the Royal Observatory at Greenwich. She found that this object, at a distance of 15600 million light-years, is receding from us at 28600 km/s. This is about 95% of the speed of light!

If we imagine the universe as a currant loaf swelling in the oven as it is baked, we get an idea of what is meant by the *Expanding Universe*. All galaxies are rushing away from us and from each other. A galaxy twice as far is rushing away at twice the speed. If this expansion has always been happening at this rate, we can calculate a time in the past when all the galaxies, or the materials of which they are made, were in one place. One idea of the history of the Universe is that it began with a 'Big Bang' about 18000000000 years ago. The Universe has been expanding ever since then and is still doing so. Most of the facts that we have observed about the Universe support this idea. For example, the ages of meteorites have been estimated by radioactive dating (p.34). Radioactive rhenium-187 decays to osmium-187. Analysis of the amounts of rhenium and osmium in meteorites suggests that their age is between 13 and 22 thousand million years.

If we believe that the Universe is expanding, we might wonder if it will continue to expand forever. Will all other galaxies eventually become so far away that they are out of sight? Or is the expansion slowing down? Just as we throw up a ball and it gradually slows down because of the pull of Earth's gravity, is the gravitational attraction between all the galaxies gradually decreasing the rate of expansion? Will expansion ever stop? Will the galaxies then begin to move towards each other as gravity overcomes the forces of the Big Bang? To answer these questions we need to know far more about the sizes and masses of objects in the Universe. Since space is so vast, we need to know more about the gases and the clouds that fill it. In particular we need to be able to look out far into space at the most distant galaxies and other objects. Since the light from these objects takes so many millions of years to reach us, we view these objects *as they were,* millions of years ago. We can look back into the history of our Universe.

To observe and measure at such distances demands new techniques. The simple optical instruments are not enough. The rest of this chapter deals with some of the modern methods of measuring the Universe.

### Interferometers

One of the difficulties in measuring the sizes of the stars is that even in the most powerful telescopes they appear only as points of light. The Sun, Moon and planets (except Pluto) appear as discs, and we can attempt to measure their diameters. But the stars are too far away. The first measurement of the diameter of a star was made by Albert Michelson, whose work on the speed of light has been described in Chapter 6. Michelson's interferometer is shown in Fig. 24. Two narrow beams of

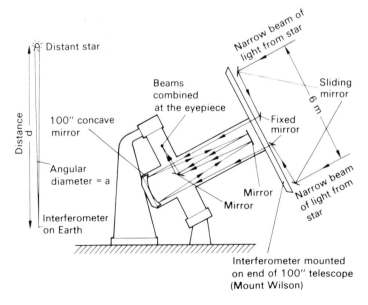

**Fig. 24.** Michelson's interferometer.

light, received by mirrors about 6m apart, are combined into one beam in the telescope. The observer does not see a single spot of light, but a number of bands of light with dark bands between them. These fringes are caused by *interference* between the light of the two beams. There is not enough space to go into the details here, but the method of using the interferometer is to vary the distance between the two outer mirrors while watching the changing pattern of fringes through the telescope. At a certain distance apart of the mirrors, the fact that the light is actually coming from *different parts* of a *disc* of light results in the pattern of fringes disappearing. The distance between the mirrors is known and from this the angular diameter (a) of the star can be calculated. The distance of the star (d) can be measured by one of the methods described in Chapter 5. We can then calculate its actual diameter. Michelson's first measurement using this instrument was the diameter of the giant star *Betelgeuse* in the constellation of Orion. He found it to be about 384 million kilometres in diameter, more than 250 times the diameter of the Sun. Since then, similar equipment has been used to measure the diameters of many other stars.

The idea of the interferometer is to combine two beams of light that have been received several metres apart. Then we can measure fine details that can not be seen in an ordinary telescope. To see the same amount of detail in an ordinary telescope would require a telescope mirror 6m in diameter. A similar idea is used in radio astronomy. A radio telescope consists of a receiving aerial or antenna, usually with a large metal reflector to collect a wide beam of radio waves. In addition, a radio receiver and amplifer are required, with some equipment for recording the signals. As the reflector dish is scanned across the sky a 'picture' is built up showing points and areas of strong radio emission. Often, the objects that are 'bright' in the radio sky can also be identified visually, using an optical telescope. Since radio waves have much longer wavelength than light, a radio telescope needs a much larger reflector dish to give the same fineness of detail as an optical telescope. Even the enormous reflector of the Arecibo Radio Observatory in Mexico, carved out of the mountains and 328m in diameter, has a resolution of only a few tenths of a minute of arc. Measurement of radio waves from distant stars and galaxies gives us a lot more knowledge about the Universe, but it is important that even finer details should be detected. Again, interferometry comes to the rescue. By using two separate radio telescopes, placed some distance apart, and by combining their signals electronically, we have an instrument that has the resolving power of a single reflector of much greater size. In another type of interferometer, such as the one constructed at Cambridge, a number of reflectors are sited along a straight line. The Cambridge telescope has 8 reflectors on a line 5 km long running east to west (Fig. 25). As the Earth rotates the east-west line is rotated too, as seen from above the North Pole. By moving some of the reflectors along rails and combining the signals from all reflectors over a period of time the array becomes equivalent to a dish 5 km in diameter. It is possible to build up a picture of the radio sky with a resolution of a few tenths of a *second* of arc. This shows details only one third the size shown by the best optical telescopes. It has revealed a hole on a hydrogen cloud that had previously not been detected. A writer in *New Scientist* remarked that this is equivalent to seeing the hole in a Polo mint at a distance of 3 kilometres.

The problem of noise in measurements was mentioned in Chapter 3. If you look at the stars through even a small telescope, the images appear to dance about, to fade and reappear. This may be due to smoke and clouds, but is also caused by convection currents in the atmosphere. Rising currents of warm air have lower density than the cooler air around them. The differences of density cause the rays of light to be bent or refracted as they pass through the atmosphere. This causes the shimmery effect on the image. It is noise, and limits the

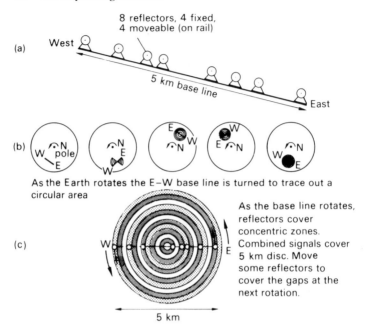

**Fig. 25.** Cambridge 5 km radiotelescope (a) Arrangement of reflectors (b) How the Earth's rotation rotates the base line (c) How the rotation gradually builds up a complete 5 km disc.

degree of resolution of the telescope. It is a limit to the precision of our telescopes. This is the reason why the world's biggest observatories are usually in remote places, away from the smoke and reflected glare of the lights of cities. Often they are on mountain peaks where the air is more often clean and still. Even there, the Earth's atmosphere can blur our vision and limit the precision of astronomical measurements. Speckle interferometry is one method used to overcome this difficulty. A series of pictures are taken electronically by a very sensitive device. Since the device is highly sensitive it can take a very large number of pictures in a reasonably short time. The information present in the pictures is fed to a computer which is programmed to combine all the pictures into one and eliminate the effects caused by the atmosphere. In other words, the computer removes the noise. It is rather as if you are listening to a distant radio station when reception is very bad. Above the crackles and whistles you can hear the same message being repeated — perhaps an SOS message. Though you can hear only a few indistinct words of each message, your knowledge of English words and grammar allows you gradually to piece together the parts you hear. In the end the whole message becomes clear — you have eliminated the effects of the noise. Speckle interferometry is used for measuring the diameter of stars and other fine astronomical detail. It has recently been used to measure the diameter of the planet Pluto. This is too far away and too small to be seen as a distinct disc. Previously it was thought that its diameter was 5900 km, but speckle interferometry has shown it to be only 3000-3600 km, roughly equal to that of the Moon. Radio telescopes too suffer from noise. Radio broadcasting is one cause of this, though certain wavelengths are set aside for radio astronomy. Noise is caused by electrical equipment being operated in the region of the observatory. Telescopes operating in the microwave range suffer interference from radar and from defective microwave ovens. Interference also comes from thunderstorms. Changes in the amount of water vapour in the atmosphere upset reception too. Another source of noise originates in the amplifier of the telescope itself. Radio signals from the Universe are very weak and high amplification is needed. A considerable amount of noise can be generated in electronic circuits by random motion of electrons. If you tune a radio set to a silent region of the waveband and then turn up the volume to its loudest, you can hear this noise as a hissing sound. When the weak signals from space are amplified in a radio telescope, this noise is amplified too. To overcome this problem, special 'low noise' amplifiers have been designed for radio telescopes. One type of low noise amplifier is cooled to below 4K (−269°C) to reduce the random motion of electrons.

The Earth's atmosphere is a nuisance to astronomers for another reason than the noise it produces. It filters out radiation of certain wavelengths so that we detect them weakly or not at all. We receive visible light and infra-red, and radio waves with wavelength between 1cm and 30m, but all other types of electromagnetic radiation is more-or-less invisible to observers on Earth. The only solution is to take our telescopes well above the atmosphere. Telescopes carried in satellites offer a way of measuring all kinds of radiation from objects in space, without any of the disturbance we suffer when making observations from Earth's surface.

*Astronomical Satellites*
One of the satellites due to be launched in the nineteen-eighties is *Hipparcos*, named after the Greek astronomer, who was one of the first to list and measure the positions of stars. His work has already been mentioned on page 52 (though there the spelling of his name is different from the version chosen by the European Space Agency). The purpose of *Hipparcos* is to continue the work begun over two thousand years ago by Hipparchus — to measure star positions and velocities. From its position in geostationary orbit, *Hipparcos* will be free from the noise created by Earth's atmosphere, and its measurements will be of extremely high precision.

The *Explorer* series of satellites includes some of the most important of the astronomical satellites. *Explorer 1*, launched in 1958, was the first American satellite to be put into orbit. It carried a geiger counter and first detected the *Van Allen belts*, the region around the Earth in which there are large concentrations of charged particles, trapped by the Earth's magnetic field. Many others of the *Explorer* series have been used for measuring conditions in the upper atmosphere and in the regions of space around Earth. They have also been used for measuring radiations from the Sun. The series includes the Small Astronomy Satellites (SAS), the first of which was called *Uhuru* and was launched in 1970. This carried equipment to measure X-rays from outer space. This was followed by the second SAS in 1973 to measure gamma rays, and the third SAS in 1975 to measure X-rays. The atmosphere absorbs X-rays so it is difficult to measure them at the Earth's surface. Earlier attempts had involved sending equipment by high-altitude balloon or by small rockets. With the development of satellites able to keep the equipment aloft for months or even years, it became possible to map the skies for sources of X-rays. Many interesting objects have been found. The first was a very bright source discovered in the constellation of the Scorpion, and named Sco X-1. This has been found to be emitting energy at an enormous rate and is thought to be a neutron star. Another object, Cygnus X-1, is possibly a black hole orbiting an ordinary star and drawing matter from it.

A new series of astronomical satellites are the High Energy Astronomical Observatories being launched by NASA. The second of these, called *Einstein*, began observations in 1980. It carries an X-ray telescope and is designed to map the X-ray sky. Since X-rays penetrate most materials, it is not easy to make lenses or mirrors for the telescope. The mirror of *Einstein*'s telescope consists of metal sleeves (Fig. 26). X-rays meet the sleeves at small angles and instead of being able to pass through they are reflected off at a small angle and sent on to the

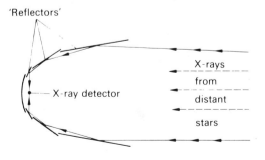

**Fig. 26.** X-ray telescope as carried on the *Einstein* satellite.

next sleeve. Gradually this is brought to a focus. The interesting thing about studying X-rays from space is that they are generated where very energetic events are occurring. Studying X-rays can lead us to an understanding of the formation of the Universe. The *Einstein* satellite has discovered more than 100 quasars. These are extremely bright but distinct objects emitting as much energy as a thousand million Suns. Closer to home, *Einstein* has detected X-rays coming from the planet Jupiter. This was the first discovery of X-rays from a planet other than Earth.

At one time it was thought that the X-ray sky was not a 'dark' one, that a 'glow' of X-rays came from all directions in space. It was suggested that the empty space between the stars and galaxies contains very large amounts of very thin gases. This gas was thought to give off X-rays and was all around us. Although the gas has *very* low density (about 1 atom per cubic centimetre), space is so vast that the total mass of this gas is enormous. Its mass *might* be enough to produce gravitational attraction on all the galaxies. If so, it would gradually slow them down. The expansion of the Universe would become slower and eventually would stop. Then the Universe would begin to collapse inward.

The improved X-ray telescopes such as those carried on *Einstein* have gradually shown that the overall X-ray 'glow' does not exist. Detailed studies of 'glowing' parts of the sky, where no bright X-ray objects have been detected, show that the 'glow' is due to numbers of distinct but weak X-ray sources far away in space. This illustrates how improvements in measuring instruments compel us to change our ideas about the Universe. The latest X-ray observations suggest that the material in space is not as dense as it was thought to be. From this evidence it seems that the Universe is likely to continue to expand indefinitely.

Evidence from radio astronomy suggests the opposite. Matter in space has the effect of slowing down radio waves as they pass through space. The longer the wavelength, the more they are slowed down. The large radio telescope at Arecibo has been used by two American astronomers, Linscott and Erkes, to measure radiation from the galaxy M87. Their measurements show that the radio waves have indeed been slowed down on their journey between M87 and Earth. Part of this effect is due to matter in M87 itself. But part could be due to matter between that galaxy and our own. If so, the amount of matter between the galaxies *is* enough to slow down the expansion of the Universe.

Measurements have shown that our Universe is expanding. With improved instruments and new techniques we are beginning to find out if expansion will continue or will eventually stop. We have no definite answer at present, but it seems likely that continuing improvements in our instruments will one day give us the precision and accuracy we need to answer this and many other questions about our Expanding Universe.

**Chapter 8**

# Measuring the Imponderable

According to the dictionary, an imponderable object is one that can not be weighed because its weight is not enough to affect a balance. While radio astronomers are looking at the largest objects in the universe, such as the quasar 3C345, which measures 78 million light years across, other scientists are measuring the sizes of objects which are too small to see. They are smaller than the wavelengths of visible light itself. If any measurement seems to be impossible to make, it is almost certain that, sooner or later, a technique or instrument will be invented that will do the job. The challenge to measure the imponderable has been one that scientists have overcome with great success.

One of the earliest successful attempts to weigh something that was thought to be unweighable was made by the mathematician, Archimedes of Sicily, in the 3rd century B.C. The king wanted a golden crown to be made. He supplied the craftsmen with a quantity of gold. When the crown was made, it had exactly the same weight as the gold that the king had supplied. But the king suspected that the craftsmen had taken part of the gold for themselves and had mixed in an equal weight of silver with the gold they used for making the crown. The king asked Archimedes to discover if this had been done.

It seemed impossible to find out the weight of gold and

the weight of the silver after they had been mixed together. The crown was not to be altered or destroyed in any way. Archimedes had this problem on his mind while he was at the bath. He noticed that water overflowed as he got into the bath. The more he immersed himself in the bath, the more water overflowed. He suddenly realised the connection between this common observation and the uncommon problem of the crown. He rushed home in great excitement and began work on solving the problem. He made a mass of pure gold having the same weight as the crown. He made a mass of pure silver, also with the same weight. He filled a bowl with water and put the gold mass into it. He measured the amount of water that overflowed. He repeated this with the mass of silver, and finally with the crown. The silver caused the most water to overflow, for its volume was greatest. Today we would say that silver has the least relative density. The gold mass caused the least overflow — it has the greatest relative density. The amount of water that overflowed when the crown was immersed was between the other two amounts. Archimedes had proved that the crown was not made of pure gold.

*Weighing the electron*
Finding the mass of the electron was but part of the investigations into atomic structure that took place toward the end of the 19th century and early in the 20th century. The first measurements were made by the British physicist J. J. Thompson in 1897. His apparatus measured the paths taken by a beam of electrons passing through a discharge tube. The tube was filled with gas at very low pressure. At that time, beams of electrons were known as cathode rays, for they appeared to come from the cathode plate of the tube. It had been established that the beams consisted of charged particles and that a magnetic field would make the electrons travel in a curved path. J. J. Thompson measured the exact effects of a magnetic field on a beam of electrons. He found that the beam was bent into a circular path as it passed through

the magnetic field. When it passed through an electrical field, its path was part of a parabola, like the path of a bullet from a gun. He then used the magnetic and electric fields *at the same time*, but directed so that one had the opposite effect to the other. By adjusting the strengths of the fields he could make the beam go in a straight line. From these results he was able to calculate the value of the ratio, $e/m$. This is the ratio between the amount of electric charge ($e$) on the electron and the mass ($m$) of an electron. These measurements could not tell him the actual values of $e$ and $m$ separately. They could only tell the ratio between them ($e/m$). So at that stage he had not succeeded in finding the actual mass of electrons. The value obtained for $e/m$ was about 180000000 coulombs of electric charge for each gram of matter. The interesting point was that whatever gas he used in the tube, the value he obtained was always the same. This suggested that electrons are the same in all kinds of matter.

Thompson's next experiment, in which he was joined by Wien, were with positive rays. These rays could be produced by making a hole in the cathode of a discharge tube. Like cathode rays, they were deflected by a magnet, though in the opposite direction. This meant that they must be positively charged, which gave them their name. Using methods similar to those used for measuring $e/m$ for electrons, Thompson and Wien measured the ratio $q/M$ for positive rays. They did this using several different gases in the tube. In this ratio, $q$ is the positive charge on the particle and $M$ is its mass. It was found that $q/M$ was *different* for different gases. For hydrogen, the value was about 100000 coulombs per gram. This compared well with results obtained from experiments on electrolysis and confirmed that when the tube was filled with hydrogen the positive rays, were charged hydrogen ions. When the tube was filled with nitrogen the value of $g/M$ was only $1/14$ of that for hydrogen. Since an atom of nitrogen was known to weigh 14 times as much as a hydrogen ion, it seemed that in this case the positive rays were charged nitrogen ions. In all the gases tried, the results showed that the positive rays were ions — atoms that had lost one or more electrons. If this is so, the charge $e$ in $e/m$ is equal in amount to charge $q$ in $q/M$. This means that we can say:

$$\frac{\text{mass of electron}}{\text{mass of hydrogen atom}} = \frac{m}{M} = \frac{q/M}{e/m} = \frac{100000}{180000000} = \frac{1}{1800}$$

This tells us that the electron is far lighter than an atom of hydrogen. Since hydrogen is the lightest atom, the electron was by far the lightest particle then known. But still J. J. Thompson had only a ratio; he still had not measured the actual weight of an electron.

The final stage was the work of the American physicist, Robert Millikan. He experimented with tiny drops of oil, suspended in air. The drops were made by spraying oil through a fine tube (Fig. 27). Spraying causes the drops to become electrically charged. He increased the amount of charging by operating an X-ray tube, which causes the molecules of air to become ionised. Some of their charge is passed to the drops. These charged drops were between two large metal plates which were themselves charged to create an electric field between them. He looked at the drops through a microscope placed at the side of the apparatus. The drops appeared as tiny specks of light moving up or

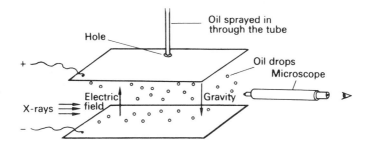

**Fig. 27.** The principle of Millikan's oil-drop experiment.

down in the field of view. The drops that were negatively charged were being pulled downward by the force of gravity, but they could be moved upward by the electrical field. Millikan could adjust the strength of the field until the upward and downward forces on one particular drop were equal. Then the drop remained stationary, neither falling nor rising. Since the force due to gravity was known, Millikan could calculate the force due to the electric field, and from that could work out how much electric charge was on the drop. He found that different drops had *different amounts* of charge. When the field was set to keep one drop stationary, other drops nearby could be seen drifting upward or downward. The interesting point was that though different drops had different charges, their charges were *all* multiples of a fixed amount of charge. The fixed amount was $1.6 \times 10^{-19}$ coulomb (0.00000000000000000016 coulomb). Some drops had $6.4 \times 10^{-19}$C (4 times the fixed amount), some had $8.0 \times 10^{-19}$C (5 times), and so on. Millikan concluded that $1.6 \times 10^{-19}$C was the charge of a single electron. Drops carried four, five or more electrons.

Millikan had found the value of $e$, the charge on an electron. It was then possible to use this value to calculate $m$, the mass of the electron. If $e/m = 180000000$ ($1.8 \times 10^8$) coulombs per gram and $e = 1.6 \times 10^{-19}$ coulombs, then:

$$m = \frac{e}{1.8 \times 10^8} = \frac{1.6 \times 10^{-19}}{1.8 \times 10^8} = 8.9 \times 10^{-28}\text{g}$$

More precise measurements give $m = 9.1 \times 10^{-28}$g.
From this we can use the ratio $m/M$ (p.85) to calculate the mass of the hydrogen nucleus, which is about 1800 times the mass of the electron. This gives us the modern value, $1.67 \times 10^{-24}$g. Since the nucleus of the hydrogen atom consists of only a single proton, this figure gives the mass of a proton too.

*The sizes of atoms and molecules*
Atoms and molecules are too small to see. We can not put a ruler beside them and measure their lengths or diameters. In that sense they are imponderable. The reason we can not see them is that they are smaller than the wavelengths of light. To measure their size we must use some rather roundabout methods. One of the earliest ways of estimating the size of molecules was to measure the *viscosity* of gases. The viscosity of a liquid such as treacle is very high. If we try to stir it, we feel a strong force in the spoon, resisting our stirring. The viscosity of water is less than that of treacle; it offers much less resistance to stirring, or the motion of an object passing through it. We might think that air or pure gases have no viscosity, but even these weakly viscous substances do offer *some* resistance to the motion of objects. The viscosity of a gas depends, among other things, on the size of its molecules. We can not go into the complicated mathematics here but measurements show that simple molecules have an average diameter of about $10^{-10}$ metres, or 0.1 nanometres. This compares with 590 nm for the wavelength of the yellow light from a sodium vapour street lamp. It is plain that visible light is of no use for viewing molecules, and yet there is an indirect way in which we can look at them. This method involves diffraction. If you have not seen the diffraction of visible light by a diffraction grating, you can obtain the effect by using a long-playing disc, as shown in Fig. 28. When light is reflected from a surface that is ruled with fine closely-spaced grooves, the reflected rays interfere with one another to produce a pattern of dark and light streaks or fringes. These are rather like those observed by Michelson on his interferometer (p.73) though they are produced in a different way. The closely-spaced grooves of an LP record act as a diffraction grating if the angle of view is low. If the light falling on the grating contains several wavelengths, as does sunlight, the interference pattern contains bands of coloured light. It gives a diffraction spectrum.

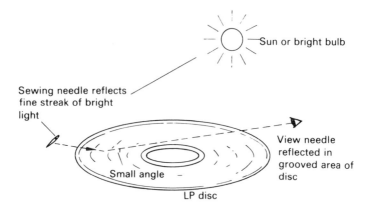

**Fig. 28.** How to see interference patterns, using an LP disc.

A crystal contains a regular arrangement of rows of molecules. We can think of it as being an extremely fine and three-dimensional diffraction grating. To get the effect, we must simply use radiation of a sufficiently short wavelength. If we use X-rays, which have wavelengths between 1 nanometre and about 0.001 nanometres, the crystal produces a complicated diffraction pattern (Fig. 29). As might be expected, the mathematics too is complicated but, by measuring a diffraction pattern, we can calculate the dimensions of the grating that produced it. In other words, we can use the diffraction pattern to tell us how the atoms and molecules are set out in the crystal. We can find distances between the rows and columns of molecules and atoms. In short, we can study the complete architecture of the crystal. X-ray crystallography, as it is called, was first devised by von Laue in 1912 and has become a powerful method for investigating the structure of matter. One of the most remarkable achievements of X-ray crystallography was the work of the English scientist, Dorothy Hodgkin, and her group of workers at Oxford. In 1955, they were the first to completely analyse the structure of an organic compound. The compound was vitamin $B_{12}$, chemically known as

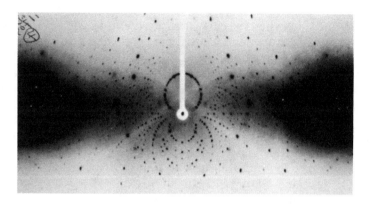

**Fig. 29.** The x-ray diffraction pattern produced by a crystal. *Photo by Birkbeck College, Department of Crystallography.*

cyanocobalamin. Deficiency of vitamin $B_{12}$ causes the fatal disease known as pernicious anaemia. The vitamin occurs naturally in foods such as meat, milk and fish. The richest source is liver. Most people obtain enough $B_{12}$ from their ordinary diet, but a few people are not able to absorb or use it easily. These people die from pernicious anaemia unless they are given vitamin $B_{12}$. Since vitamin $B_{12}$ is destroyed by cooking, the only cure for the disease was to eat large quantities of raw minced liver every day. The large quantities were necessary because the person could use only a small fraction of the amount supplied. Once the chemical structure of the vitamin had been discovered by X-ray crystallography, it became easier to synthesise the vitamin artificially. Nowadays a person with pernicious anaemia needs only a monthly injection of a solution of the pure vitamin. Since the vitamin $B_{12}$ is injected directly into the blood stream (instead of being taken in by mouth) only small amounts need be given. This measurement of the arrangement of atoms in the molecules of vitamin $B_{12}$ has resulted in the saving of countless lives and has produced a simple cure for this previously unpleasant and serious disease.

*Is the atom solid?*

At the time when the early measurements of the sizes and charges of atoms and atomic particles were being made, there were conflicting ideas about the structure of the atom itself. The older idea was that an atom was simply a spherical body, with a diameter about $10^{-10}$ metres. The more recent idea was that the atom consisted of a very dense central region, the nucleus, surrounded by a less dense region consisting of a cloud of electrons in orbits around the nucleus. Experiments by Ernest Rutherford at Cambridge in 1911 showed that the second idea was the right one. In one experiment he exposed a thin film of aluminium to a beam of alpha particles from a radium source. Most of the particles passed straight through the film and struck the screen. The screen was coated with zinc sulphide, which emitted a flash of light each time it was struck by an alpha particle. A small circular area was seen glowing at the centre of the screen. There were also occasional flashes at the edge of the screen. A few of the alpha-particles did not pass straight through the film but were deflected by as much as 40° from their path. The fact that most particles went straight through suggested that the atoms of aluminium consisted mainly of empty space. Alpha particles are positively charged and those few that happened to pass close by the positively charged nucleus of an aluminium atom were strongly deflected. From the numbers of deflected particles that he observed, Rutherford estimated that the diameter of the nucleus of the atom was in the order of $10^{-14}$m. This is only about one *ten thousandth* of the diameter of the whole atom. Truly the atom does consist mainly of empty space.

*The electron microscope*

One of the still unexplained things about light is that although we usually think of it as having the features of *waves*, it also shows some of the properties of *particles*. Conversely, a beam of electrons, which we usually think of as a beam of particles, also shows many of the properties of *waves*. A beam of electrons accelerated under high voltage has the properties of radiation, with a wavelength as short as that of X-rays. This means that a microscope that uses electron beams instead of light can be used to view extremely small objects. The electron microscope can magnify over 15 million times, so we need only the tiniest piece of material to look at. Electron microscopes allow us to measure the sizes of small living organisms such as viruses and the molecules of which they are made. High-resolution electron microscopes allow us to look at the atoms themselves. We can examine a crystal of a mineral and view its patterns of atoms. We can measure the distances between them with an accuracy of $10^{-11}$ metres, or 0.00000001 millimetres. In this way we gain knowledge of the structure of materials and can begin to understand how their structure gives them their special properties.

**Chapter 9**

# Measuring the Indefinable

It is easy to say what we mean by length, mass, heat or an electric current, so it is easy to decide exactly how we will set about measuring these quantities. But how do you set about measuring the loudness of a noise? It is important to be able to measure this, for noise can be a nuisance. We have laws to control the amount of noise that a lorry is allowed to produce, or that an aircraft may make when taking off. We can not enforce those laws unless we have some way of measuring loudness. It is also important to be able to measure noise when designing buildings, concert-halls and recording studios. The unit in which noise is measured is the *decibel*. The decibel scale is based on the *power* of the source of sound.

The eye is reasonably good at judging relative lengths. For example, we can see that one object is about twice as long as another. The ear works differently. A sound that has twice the power does not sound twice as loud. The smallest difference of power that your ear can detect is about 1 decibel — which is equivalent to an increase of power of about 25%. Note the phrase '*difference* of power' in the last sentence. The decibel is not a unit such as the kilogram in which 1 kilogram of matter is a fixed amount of substance. The decibel compares the power of one sound with the power of *another sound*. Ten decibels is a power *difference* of about 1000% or 10 times. The sound of someone whispering is about 10 decibels *louder than* the sound of leaves rustling in the wind. The sound of a heavy lorry is about 10 decibels *louder than* that of a motor scooter. The actual difference in power between rustling leaves and whispering is only a few microwatts. Between a motor scooter and a lorry it is several tens of watts. Yet both differences are rated at about 10 decibels. This scale suits us well because of the way our ear works — as we go up the scale of loudness we need to take bigger and bigger steps (as measured by power) to get equal steps of loudness (as heard by ear). One serious result of this is that the power of a very loud sound is considerably greater than that of a sound that is not *quite* so loud. The slightly louder sound does not *sound* that much louder but, even so, it may be so much more *powerful* that it damages the ear. Exposure to excessively powerful sounds (whether they sound loud or not!) causes permanent damage to hearing and leads to deafness in later life. This is another reason why we need to measure loudness.

To measure loudness we use a microphone connected to an amplifier. The amplifier measures the power being received by the microphone. The amplifier operates a meter that indicates the power directly, in decibels. These meters are usually calibrated so that the zero on the scale corresponds with the quietest sound that the average human ear can detect. At the top end of the scale (around 130 decibels) we have the noise made by a riveting gun, a noise that is ten million million times more powerful than the quietest noise we can hear (Fig. 30). In spite of all the measurements, there still remains the fact that different people can have different opinions about the loudness of the same noise. A noise we do not like, such as the neighbour's motor mower, sounds much louder to us than it does to him. A countryman visiting the city may find the unfamiliar traffic noise unpleasantly loud, while the person who works in the city every day hardly notices them. This makes it all the more important that we have a yardstick, the decibel, for measuring noise.

```
  0   the quietest sound possible
 10   rustling leaves
 20   whispering
 30   a ticking clock at 1m
 40   a quiet room with an open fire burning
 50   a quiet street
 60   a conversation
 70   a busy street
 80   a motor-cycle
 90   a heavy lorry
100   a jet aeroplane at 200m
110   a jet aeroplane at 100m
120   a pneumatic drill at 5m
130
```

**Fig. 30.** The decibel scale.


*Measuring the mind*

Although we can measure the power of noise in watts, we get into difficulties when we try to measure its mental effect — how loud a particular noise *appears* to the person hearing it. The measuring of mental states and activities is important to people such as psychologists, doctors and teachers. The chief difficulty is to define exactly what you wish to measure. In 1905, Alfred Binet, was asked to design a test for measuring the mental abilities of French schoolchildren. The test that he and Herbert Simon invented, often called an 'intelligence test', has since been revised by American workers and is now widely used. Exactly what is meant by 'intelligence' and exactly what this test measures are both almost impossible to define. Yet, despite this, we find the test a useful one.

The test consists of a set of questions, simple problems, and exercises in reasoning, suited to children of different age groups. The tests themselves are tested on large numbers of children, to see just what can be expected of the average child at each age. There are tests for seven-year-olds, tests for eight-year-olds and so on over the range of school ages. We might use the tests on a girl who is, for example, nine years old. If she can answer the nine-year-old test satisfactorily, she is given the ten-year-old test. If she can manage this too, she is tried with the eleven-year-old test. She may be successful with this too but do poorly in the twelve-year-old test. We have said that the girl is nine years old: this is her *physical* age. The test shows that her *mental* age is eleven years. Mentally, she has the ability of an average eleven-year-old child. To compare abilities of children of different physical ages we use the *intelligence quotient*, usually shortened to I.Q. This is calculated as follows:

$$I.Q. = \frac{\text{mental age}}{\text{physical age}} \times 100$$

In this example of the girl, her I.Q. is

$$\frac{11}{9} \times 100 = 122$$

Measurement of I.Q.'s of a large number of children of the same age group shows the average I.Q. is 100. This must be so, for the average child should, by definition, be able to pass the test for his or her average age. About one sixth of the age-group have I.Q.'s greater than 116 and very few indeed are higher than 140. The girl of our example with an I.Q. of 122, is one of the more intelligent of her age. If half of the children are of average intelligence or more, it follows that the other half are of *less* than average intelligence. About one sixth have an I.Q. of 84 or less, and a very few are less than 70.

The Binet test measures 'something' but what does it really measure? When we know a person has an I.Q. of 122, what does this tell us about this person? Perhaps such test *on its own* can tell us very little. The Binet test is perhaps too general. We need other tests that separately measure special abilities such as verbal ability, memory, and reasoning skills.

Perhaps even more difficult than measuring abilities is the measuring of personality. Can we measure pride, perseverance or criminality? Psychologists believe that we can, and many measurement tests have been devised. As an example, let us look at one of the tests for measuring *suggestibility*.

Some people are much more likely than others to accept ideas suggested to them. For example such people may more readily believe what they see on the TV commercials. These people are very suggestible. Others are only slightly suggestible. The measurement of suggestibility could be very interesting to the producers of TV advertisements. They would be interested to know which types of people are most suggestible. Several tests have been devised, including the following one. The persons being tested are asked a number of questions dealing with topical subjects. The questions are not factual ones but ones on which people can be expected to have fairly strong opinions, and on which people can have widely *differing* opinions. They might include such subjects as conservation and pollution, the use of nuclear weapons, and the policies of the government of the day. They write their answers to these questions and are told to express their own opinions freely in their answers. Their answers are then collected. Next they are given a larger set of questions of similar kind. Among this second set are most of the questions of the first set. Before they answer the second set of questions, the people are told that an important person has also answered these questions. They are told that the important person is the Prime Minister, or some other well-known person of authority. It is found that many people give *different* answers to the *same* questions on the second occasion. Generally their answers are more in line with answers that they think the authority would probably give. Suggestible people alter their opinions to fit what they think the person of authority would think. By seeing how many answers are changed and how much these are changed, investigators can get a measure of a person's suggestibility.

*Measuring time*

It may seem strange to include time in this chapter. Time is, after all, familiar to everyone. We all know what a week is; we even use time for measuring distances (p.2) Furthermore, we can measure time so precisely, to a millionth of a second. Yet no one really knows what time is. There have been many ideas and discussions on the nature of time, but time is still undefined. This is why it is included in this chapter, and just as we have not tried to define intelligence and suggestibility before working out ways of measuring them, we will not make any attempt to define time. We will just tell the story of the measurement of time through the ages.

There has always been a great need to measure time. As might be expected, the natural yardsticks were the first to be established. The regular coming of the seasons, with regular times for sowing seeds and harvesting, soon led to the acceptance of the *year* as an important measure of time. The day is another such natural yardstick. The *month* is a sub-division of the year that is of convenient length for many purposes. It is based on the period taken for the Moon to pass once in orbit around the Earth. The period is just over 27 days. The exact length of a month depends on exactly what you mean by 'a month'. For use in the calendar, the people of most countries divide the year into twelve months between 28 and 31 days in length. This is the Gregorian calendar, named after Pope Gregory XIII, who introduced it in 1582. It has about the same length as the solar year, the time taken for the Earth to orbit once around the Sun. Once again we have a unit being subdivided into twelfths (see p.5). In Moslem countries the month is a synodic month, beginning in the evening soon after sunset with the first sighting of the thin crescent of the new moon. The Moslem calendar therefore has about 13 months in the year, though there are 12 months in the Moslem year. As the years pass, the Moslem feasts and holy days fall earlier and earlier in the Gregorian Year. Because each month begins at a moonrise, which is at about six o'clock in the evening, it is customary in some Moslem countries for the day to begin at 6 p.m., rather than at midnight. The *week* has been used as a measure of time from as long ago as there are written records. It does not correspond to any natural yardstick, though it seems to be accepted as a handy length of time on which to base social and working life. Possibly it began as a quarter of a month.

The shortest natural period is the *day*. Shorter periods than this are obtained by sub-divisions of the day. In the past, the *hour* has not always had a fixed length. The period of daylight, from sunrise to sunset, was divided into a number of hours but, since the length of daylight varies with the season, the length of the hour varied too. This did not matter in the countries of the Mediterranean and Middle East where these systems were first used, for the times of sunrise and sunset do not vary greatly during the year. In more northern latitudes the times vary a lot and made accurate time-keeping almost impossible. The 24-hour day was soon adopted. Here again we have the now familiar idea of division into twelfths, twelve hours for the day and twelve for the night. One of the earliest ways of measuring hourly times was the sun dial. As early as 1600 B.C. the Egyptians had used tall pointed stones (obelisks) to cast a shadow on the ground. The position of the shadow indicated the time of day. Another form of Egyptian sun-dial, the shadow-stick, is shown in Fig. 31. This measures the Sun's altitude in the sky by the length of the shadow cast on the stick. The stick is marked to show the hours. The more common type of sundial has a pointer or *gnomon* that casts a shadow on a dial (p.103). This kind of dial was used in many parts of the world and there are still plenty to be seen in gardens and on old buildings. A sundial normally indicates *local* time, though it can be marked to show some standard time, such as Greenwich Mean Time, if required. The orbit of the Earth around the Sun is an ellipse, not a perfect circle, and the Earth's axis is tilted at about 23½°, so the hours measured by a sundial are

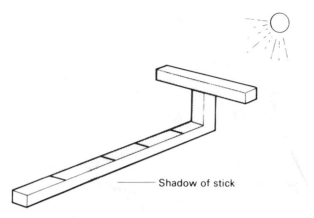

**Fig. 31.** Egyptian shadow-stick.

not constant in length throughout the year. We prefer to have hours of constant length, averaging out the variations in the apparent motion of the Sun over a whole year. Our time is based on this average (or *mean*) motion of the Sun, giving us our standard time, Greenwich *Mean* Time. The time measured by a sundial differs from mean time by an amount called the *equation of time* (p.103). We calculate mean time, by finding the equation of time for any given day of the year and adding this to our sundial time. The inaccuracy due to the equation of time does not matter much for everyday purposes, but the fact that the sundial does not work at night or in cloudy conditions is a big draw-back. The astrolabe (p.67) could be used on clear nights and there were several other simple clocks in general use. The clocks shown in Figs. 32 and 33 all measure *elapsed time*, that is, the length of time that has passed since they started to run. Obviously the candle and string were highly inaccurate, but useful enough for everyday purposes. The water-clock, or *clepsydra*, dates back from 1400 B.C. The vessel is narrower toward the bottom, to allow for the fact that the water runs out more slowly as the level falls. The clepsydra was widely used for thousands of years. One of its disadvan-

**Fig. 32.** An Egyptian Clypsedra or water clock. *Photo by Michael Holford. Reproduced by kind permission of the Science Museum, London.*

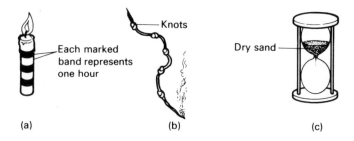

**Fig. 33.** Primitive timepieces (a) marked candle (b) smouldering knotted string (c) sand-glass.

tages is that temperature effects the viscosity of water, so the water runs out more quickly in warm weather. In some of the later forms of clepsydra there was a float in the vessel, attached to a cord. As the water level fell the float fell with it, pulling on the cord. This made a pointer rotate to indicate the time on a clock face. A clepsydra was used by Galileo in his investigations into the motion of falling objects. It is hard to imagine how such a clock could be used to time such a rapid action, but Galileo overcame the problem in an ingenious way. Instead of working with freely falling objects, he used a ball rolling down a gentle slope . The laws governing the acceleration are the same as for freely falling objects, but the ball takes much longer to run down a track than to fall through the air. The times were long enough to be measured with reasonable accuracy by a clepsydra. Galileo collected the water running out from the clepsydra while the ball rolled a measured distance down the slope. He was able to show that the distance rolled is proportional to the square of the time taken. If, for example, the ball rolls for twice the time, it travels four times as far.

Although sundials, astrolabes, water-clocks, and hour-glasses had been good enough as time-keepers for about 300 years, the increasing interest in exploration of the Earth, and in scientific investigations generally, made it important to have clocks that were more precise and accurate. Mechanical clocks of various designs had been built since the Dark Ages (5th to 8th centuries A.D.). They were usually driven by a falling weight. As the weight falls, the hands (or more usually a single 'hour' hand) are turned around a dial. The rate at which the mechanism can operate is controlled by an *escapement*. The escapement is the time-keeping part of the clock. The clock's accuracy depends on this, and much of the improvement in clock design lies on designing better and better escapements. For example, the medieval clocks used a *foliot* (Fig. 34). The positions of the weights could be altered to alter the speed of the clock. Because the

## Project 4: Make a Sundial

Timepieces like this one were in use for thousands of years.

(1) *Make the dial plate.*

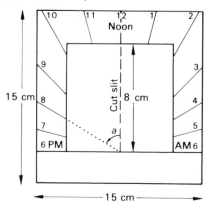

(2) *Mark out the angles:*

(a) Find the sine of your latitude angle eg. if your latitude is 53 ° N, sine of 53 ° is 0.804.

(b) Multiply each of the figures below by this sine:

| Times | | Value |
|-------|------|-------|
| 11am | 1pm | 0.268 |
| 10am | 2pm | 0.577 |
| 9am | 3pm | 1.000 |
| 8am | 4pm | 1.732 |
| 7am | 5pm | 3.732 |

eg. for 8am (and 4pm), 0.804 x 1.732 = 1.393

(c) Find the angle that has this value as its tangent (use tables or calculator):

eg. angle for which tangent is 1.393 is 54°. Angle *a* in the drawing is 54°. Angles for 6 am and 6 pm are 90°.

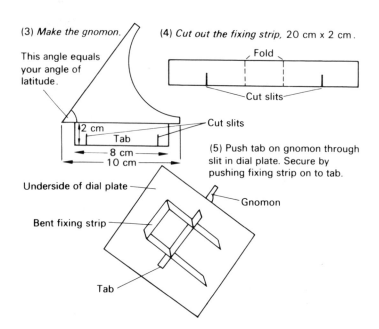

(3) *Make the gnomon.*

This angle equals your angle of latitude.

2 cm
Tab
8 cm
10 cm

(4) *Cut out the fixing strip,* 20 cm x 2 cm.

Fold

Cut slits

Cut slits

(5) Push tab on gnomon through slit in dial plate. Secure by pushing fixing strip on to tab.

Underside of dial plate

Gnomon

Bent fixing strip

Tab

(6) *Finding the meridian* (local N–S line).

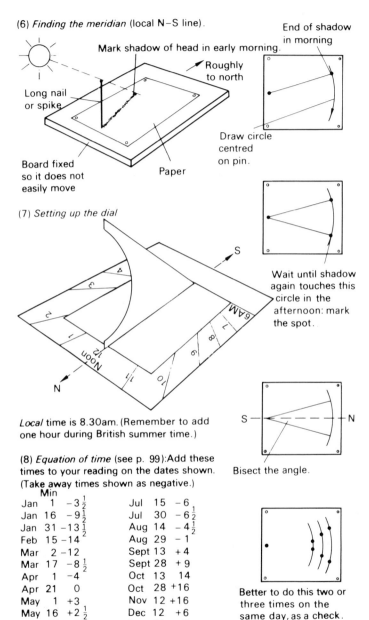

Mark shadow of head in early morning.

Roughly to north

Long nail or spike

Board fixed so it does not easily move

Paper

End of shadow in morning

Draw circle centred on pin.

(7) *Setting up the dial*

S

Wait until shadow again touches this circle in the afternoon: mark the spot.

N

*Local* time is 8.30am. (Remember to add one hour during British summer time.)

S — N

Bisect the angle.

(8) *Equation of time* (see p. 99): Add these times to your reading on the dates shown. (Take away times shown as negative.)

| Min | | | |
|---|---|---|---|
| Jan 1 | $-3\frac{1}{2}$ | Jul 15 | $-6$ |
| Jan 16 | $-9\frac{1}{2}$ | Jul 30 | $-6\frac{1}{2}$ |
| Jan 31 | $-13\frac{1}{2}$ | Aug 14 | $-4\frac{1}{2}$ |
| Feb 15 | $-14$ | Aug 29 | $-1$ |
| Mar 2 | $-12$ | Sept 13 | $+4$ |
| Mar 17 | $-8\frac{1}{2}$ | Sept 28 | $+9$ |
| Apr 1 | $-4$ | Oct 13 | $14$ |
| Apr 21 | $0$ | Oct 28 | $+16$ |
| May 1 | $+3$ | Nov 12 | $+16$ |
| May 16 | $+2\frac{1}{2}$ | Dec 12 | $+6$ |
| June 21 | $0$ | Dec 21 | $0$ |
| June 30 | $-3\frac{1}{2}$ | Dec 27 | $-1$ |

Better to do this two or three times on the same day, as a check.

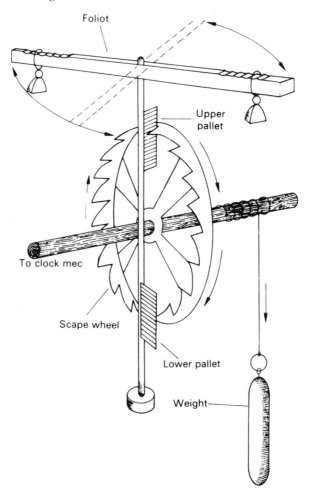

**Fig. 34.** A verge escapement: as the foliot rotates to the dotted position, the upper pallet releases a tooth. The wheel moves round but the lower pallet turns and stops it moving more than 1 tooth space. Then the foliot turns back: the lower tooth is released and the next tooth held by the upper pallet.

length of hours depended on the length of day (p.98) the speed of the clock had to be altered quite often. Such a mechanism was very inaccurate: clocks might gain or lose 30 minutes or more a day.

The next development in time-keeping came with the introduction of the pendulum. The idea had been thought of by the great artist, mathematician and musician, Leonardo da Vinci in the 15th century. It seems that he did not put this idea into practice. A century later Galileo noted the regular swinging of the chandelier in the cathedral at Pisa. To measure its regularity he compared its period of swinging with the period of another regular action — his own pulse. He suggested that a pendulum would be excellent for controlling the escapement of a clock, for he recognized that the period of swing of the pendulum is not affected by the size of the swing. It depends only on the length of the pendulum. Galileo was then too old to build the clock himself. His son Vincenzio attempted to build one, but died before it could be completed.

Another century passed before the first really successful pendulum clock was built by the Dutch physicist, Christian Huygens. In 1656 he made a clock in which the pendulum swung between two curved metal strips (Fig. 35). Although the period of a pendulum is not much

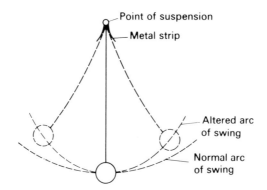

**Fig. 35.** Huygen's pendulum.

## Project 5: **Make a pendulum clock**

This cardboard clock does not keep accurate time, and it does not run for long. But it does demonstrate how an escapement works and how a pendulum is used to control the rate of working. These two mechanisms revolutionised timekeeping and made possible countless scientific measurements and discoveries.

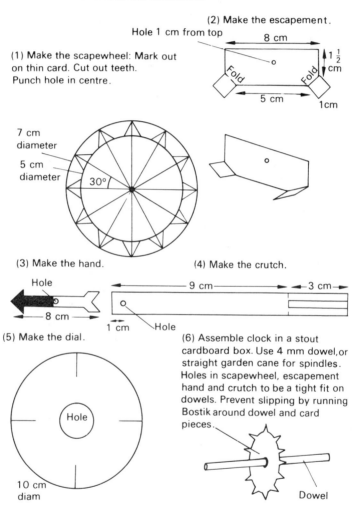

(2) Make the escapement.

Hole 1 cm from top

8 cm

(1) Make the scapewheel: Mark out on thin card. Cut out teeth. Punch hole in centre.

$1\frac{1}{2}$ cm

5 cm

1cm

7 cm diameter

5 cm diameter

30°

(3) Make the hand.

(4) Make the crutch.

Hole

9 cm        3 cm

8 cm

1 cm    Hole

(5) Make the dial.

Hole

10 cm diam

(6) Assemble clock in a stout cardboard box. Use 4 mm dowel, or straight garden cane for spindles. Holes in scapewheel, escapement hand and crutch to be a tight fit on dowels. Prevent slipping by running Bostik around dowel and card pieces.

Dowel

Make the holes in the box a *loose* fit.

As the pendulum swings, the escapement rocks to and fro, letting the scape wheel turn *one tooth at a time*. Weight gives force to turn the scape wheel and hand, and helps keep the pendulum swinging.

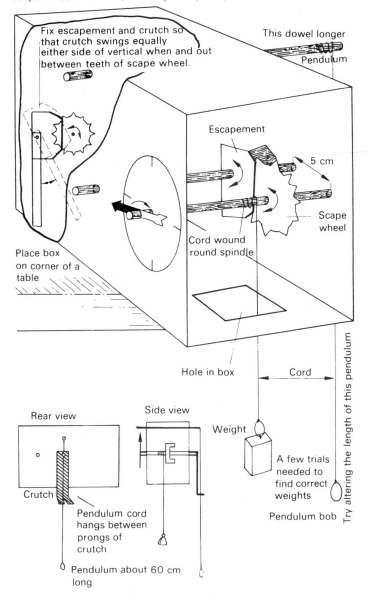

Fix escapement and crutch so that crutch swings equally either side of vertical when and out between teeth of scape wheel.

This dowel longer

Pendulum

Escapement

5 cm

Scape wheel

Cord wound round spindle

Place box on corner of a table

Hole in box

Cord

Rear view

Side view

Weight

Crutch

Pendulum cord hangs between prongs of crutch

Pendulum about 60 cm long

A few trials needed to find correct weights

Pendulum bob

Try altering the length of this pendulum

affected by the size of swing, there is a slight effect. Huygens knew that the period of swing is constant only if the size of the swing is small. A widely swinging pendulum swings more slowly. The metal strips acted to shorten the pendulum as it swung outward. The bigger its swing the more it is shortened. If the strips are correctly curved the period of swing can be kept constant for all sizes of swing within given limits.

One development in clock-making was the use of the balance wheel instead of the pendulum. The other was the use of a spring as a source of power instead of a falling weight. The balance wheel is very suitable for portable clocks and watches. The spring is essential for watches and useful in other types of clocks. It has the disadvantage that it produces a very strong force when newly wound and gradually less force as it unwinds. The fusee (Fig. 36) was an invention designed to produce a constant force from a spring. As the spring unwound, the leverage of the chain became gradually greater, so the power to the watch was held constant. Even so, watches were usually no more accurate than half-an-hour a day.

With gradual improvements in design, clocks and watches became more precise. But as we saw in Chapter 4 (p.42) a precise watch is of little use if it is not set accurately. Today we are used to being able to obtain accurate GMT by telephoning the speaking clock, by listening to the time signal on the radio, or by reading the time to the nearest second on Teletext. In the 17th century there was no more accurate way of setting a watch than using the nearest sundial! This lead to complications because of the equation of time (p.99). Although clocks were precise they were very often inaccurate. Even if they knew about the equation of time, most people found it hard to adjust their sundial times correctly.

The invention of really accurate clocks, or chronometers as they are often called, was spurred on by the needs of navigators at sea. As was explained on p.28, a navigator can not accurately find his longtitude unless he knows what the time is at Greenwich. He needs a clock

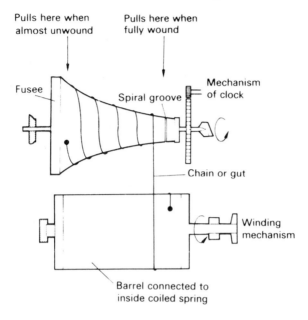

Pulls here when
almost unwound

Pulls here when
fully wound

Fusee

Spiral groove

Mechanism
of clock

Chain or gut

Winding
mechanism

Barrel connected to
inside coiled spring

**Fig. 36.** The fusee mechanism for spring-powered clocks.

that can be set to GMT at the beginning of a voyage. This could be a voyage of several months and the clock must keep accurate time for the whole voyage.

There was one alternative method, the *method of lunar distances*, favoured by Nevil Maskelyne, the fifth Astronomer Royal. In this method the position of the Moon was determined by measuring its angular distance from several bright stars. Tables were published to show where the Moon would be in the sky at different 3-hourly Greenwich times and at different dates during the year. Since the Moon was visible from all parts of the world, it was possible to tell the Greenwich time from any part of the world. Although this method needed no expensive equipment, other than a sextant, the tables were not completely accurate. If the Moon was not visible, no reading could be taken. This method was in use from 1770 until 1907, when the tables were no longer published.

It was so important that ships should be able to find their longtitude that, in 1714, the British government offered a reward to anyone who could devise a suitable method. The reward was up to £20,000, depending on how precise the method proved to be. This was a considerable sum of money in those days, equivalent to several millions of pounds by today's values. There were several contestants, but the winner of the prize was the English instrument maker, John Harrison. His first chronometer was completed in 1735 and taken on a voyage to Lisbon in 1736. Such a clock must not only be accurate when on dry land but must not be affected by the rolling and pitching of a ship at sea. The test was successful, but Harrison continued to make several more chronometers of improved design. One of the most accurate of these ran with an error of only 4½ seconds in 10 weeks. As a result of his work, explorers such as Cook were able to sail the world, charting new shores with an accuracy never possible before. In 1777, Harrison was finally paid the reward. Several of his chronometers are still working and keeping accurate time today (Fig. 37).

In recent years the development of electronics has revolutionised the measurement of time. The power to drive a clock can be provided by electricity. We can display the time by hands moving over a dial or by a digital display.

The regulation of the time piece can be done in various ways. One of these uses a timing tuning-fork which, like all tuning forks, has the property of vibrating at a constant frequency. Energy is supplied to the fork by an electromagnet, to keep it vibrating continuously. Its vibrations are detected electromagnetically and used to control a small motor, that drives the hands at constant speed. In electronic watches and clocks a quartz crystal is made to vibrate electronically. A crystal can be cut so as to vibrate at a precisely defined frequency. Such crystals are also used to control the frequency of a radio transmitter. Those who enjoy flying radio-controlled model aeroplanes will know that the frequency at which the

**Fig. 37.** One of Harrison's chronometers. *Photo reproduced by kind permission of the National Maritime Musuem.*

transmitter and receiver work depend upon the frequencies of the crystals plugged into the circuits. As the crystal of an electronic watch vibrates, its vibrations are counted. A commonly used standard crystal is made to vibrate 4194304 times per second. This may seem a strange number to choose, but it happens to be $2^{22}$ (2 multiplied by itself twenty-two times). It is easy to make an electronic circuit that will give out one electrical pulse for every 4194304 pulses it receives from the crystal. This provides 1 pulse per second to drive the clock. Such watches can be very cheaply made, with a precision of a hundredth of a second and an error of less than 15 seconds per month. For most purposes we need nothing better. More precise quartz crystal clocks can be made and, since 1942, have been used instead of pendulum clocks at Greenwich. The crystal is made with the highest possible degree of precision and is kept at constant temperature in a sealed container. The accuracy of the best of them is about 0.0003 seconds *a year*. Unfortunately, even the most accurate of quartz crystals changes its properties with age. The quartz crystal clocks need to be reset occasionally against our most accurate type of clock, the atomic clock. As explained on p.45 this relies on the vibrations of atoms of caesium, which do *not* change their properties with age. Since 1972 atomic clocks have been used to maintain the time system of the world. Long gone are the days when you set your watch against a sundial! Now it is possible to receive precise time signals almost anywhere in the world. Most people rely on receiving the 'pips' by radio or by telephone, but there is a special service for scientists and others for whom highly precise timing is essential. Time signals from atomic clocks are continuously broadcast by radio from certain stations. These signals can be picked up by special clocks that have a radio receiver built into them. The radio signals drive these clocks so that they keep exactly in step with the atomic clock. By this means astronomers in different parts of the world can be sure that their clocks are not only running at precisely the same rate, but that they all show exactly the same time. The world's seismographic stations can operate on the same time system. They can record the exact times that tremors of earthquakes reach each station, and so the position of the disturbances can be precisely located. So much of scientific measurement depends on precise measurement of time. It is strange that time, which is so important to us and which we have learned to measure with such high precision, still remains so mysteriously indefinable.

**Chapter 10**

# Measurement and Science

'When you cannot measure . . . your knowledge is of a meagre and unsatisfactory kind'. *Lord Kelvin*

According to popular belief, Galileo Galilei, in 1591, dropped two cannon-balls from the top of the leaning tower of Piza and in a few seconds disproved an idea that had been firmly held for 1800 years. Aristotle had said that heavier bodies fall faster than light ones. Galileo took two cannon-balls of greatly different mass and dropped them at the same instant, so demonstrating that they hit the ground at the same instant. Galileo showed that Aristotle was wrong. His experiment measured the falling times of the two balls by comparing them with each other. They were proved to be exactly equal.

It is possible that this story is not true. Galileo's own accounts of his work refer only to his measurements of the motion of cannon-balls along a gently-sloping beam, as described in Chapter 9. By this series of measurements he established for the first time the exact relationship between the time of fall and the distance travelled. When we measure something, and can express it in figures, we can then go on to induce the scientific laws that govern it. Galileo's work also led on to other studies of motion by later workers.

Whether there is any truth in the story of the leaning

tower of Pisa or not, the idea of the experiment has been modernised several times over the years. In the 19th century a more startling demonstration was the guinea and feather experiment. A golden guinea coin and a downy feather are enclosed in a long glass tube. When the tube is turned on end the coin naturally falls faster than the feather, because the feather has to overcome a far greater air resistance. After the tube has been connected to a vacuum pump and the air has been removed from it, the experiment is tried again. This time the guinea and feather fall at the same rate. The experiment was repeated in even more modern form by *Apollo* astronauts on the surface of the Moon, using an American dollar coin and the feather of an American eagle.

When Archimedes investigated the composition of the King's crown, as described in Chapter 8, he was simply doing what the King had asked. Yet the measurements he made and the thoughts he had on this subject led to the ideas of relative density. They also led to new ideas about the forces on floating objects. We still use these ideas today. One of the most important is Archimedes' Principle. By measuring certain quantities (such as mass and volume) we are able to express other quantities that are not directly measurable (such as density). New ideas lead to new thoughts and on to even newer ideas. Precise measurement is an essential part of this process. Hipparchus (p.52), in the 2nd century B.C., measured and listed the positions of stars. He noticed that his measurements differed from those made by earlier astronomers. This led him to the discovery of *precession.* As the Earth moves through space its axis slowly turns, like the axis of a wobbling spinning top. This turning action takes 26000 years to complete but can be detected by comparing records of star positions taken several hundred years apart. Hipparchus discovered precession as a result of his accurate measurements. The knowledge that precession occurs helped astronomers coming after Hipparchus to improve the accuracy of their own observations.

In the centuries following, the distances of the Sun and planets were measured (Ch.5). When these distances were set down on paper an interesting fact was discovered. This was first discovered in the 18th century by the German mathematician Johann Daniel Titius, and later published by another German, the astronomer Johann Bode. It has since become known as the Titius-Bode law. The law may be set out like this:

(1) Begin with this series
    of numbers                    0, 3, 6, 12, 24, 48, 96
    (each is double the one before
    it, except the first two)

(2) Add 4 each            4, 7, 10, 16, 28, 52, 100

(3) This gives the approximate
    distances between each planet
    and the Sun, taking Saturn as
    100

| | |
|---|---|
| Mercury | 4 |
| Venus | 7 |
| Earth | 10 |
| Mars | 16 |
| —— | 28 |
| Jupiter | 52 |
| Saturn | 100 |

The law fitted facts well, but there was a gap at the 5th place. It seemed that there ought to be a planet there, but such a planet had not been discovered at that date. For a long time astronomers searched for planets to fill the gap. Eventually the largest of the asteroids, Ceres, was discovered. Since then we have found countless other such 'minor planets' orbiting in a band between the orbits of Mars and Jupiter. In the meantime Herschel had discovered Uranus and this planet too fitted the law, at distance '196' on the scale. In the case of Uranus the fit was not as good as that of the other planets. Later, when

Neptune and Pluto were discovered it was found that they did not fit the law at all, so nowadays we can not really consider this to be a law. However, it had served its purpose in leading to the discovery of the asteroid belt.

The discovery of Neptune was also the result of careful measurement. After Uranus had been discovered, its motion was found to be irregular. It seemed that it was changing in speed as it orbited the Sun. This could be explained by supposing that an undiscovered planet lay beyond Neptune. Sometimes the gravitational attraction of this planet would be able to hold Uranus back, at other times it would make it move faster. Mathematicians were able to calculate where this undiscovered planet should be found in the sky. Later, this part of the sky was searched, and the planet was found. It was given the name Neptune. As was the case with Uranus, astronomers had noticed it before then, but thought it to be a faint star.

*Measurement in Biology*

People often think that measurement is not important in biology. Some people even describe biology as an 'inexact science'. Certainly some biological activities do not involve measurement, for example, cataloguing the animals found living in an area of moorland, or describing the behaviour of birds at a bird-table. These activities are *descriptive*. But this does not mean that biology as a science is wholly descriptive too. Early astronomers catalogued the stars and described their behaviour — when they rose, when they set, how brightly they shone, and if their brightness periodically varied. Gradually, as more and more information about the stars was gathered, and as special instruments (astrolabe, sextant, telescope) were invented, astronomers turned their attention from description to measurement. No one could claim that modern astronomy is an inexact science. It depends on measurements of the highest possible precision and accuracy. The greater part of this book is about astronomical measuring.

In the same way, biology began as a descriptive science but has gradually become more and more an exact science, based firmly on information obtained by making measurements. Perhaps it has lagged behind astronomy, physics and chemistry but today it has caught up. Many of the instruments and techniques described in this book, for example, the electron microscope, X-ray diffraction, the pH meter, the colorimeter, the measuring of decibels and all the instruments devised for precise measurements of length, mass, and time are put to constant use by to-day's biologists.

One of the first biologists to make use of measurement was William Harvey. In the 17th century it was known that the human heart pumped out blood into the tissues of the body, but it was not known what happened to it when it reached the tissues. It seemed to disappear. It was thought possible that the heart was continually being supplied with newly-made blood and that this blood was disappearing when it reached the tissues. A descriptive biologist might have been content with this explanation, but Harvey made some measurements. He found that the heart pumps about 60g of blood at each beat. If it beats 72 times a minute, the amount of blood pumped is almost 4.4kg. In an hour the heart pumps out 264kg, which is over *three times* the weight of the whole body. Harvey saw that it was impossible for the body to make and destroy three times its own weight of blood every hour. He correctly concluded that the blood pumped through the heart is always the *same* blood *circulating* around the body. There must be some way for the blood that is pumped into the arteries to travel to the veins and return to the heart again. Microscopes had not been invented in those days, so Harvey could not see the capillaries which we today know are the connecting link between arteries and veins. He could only say that there must be tubes so fine as to be invisible to the eye. In fact, the body contains about 5kg of blood, which takes 1 minute to circulate.

Shortly after this a number of people, including

Anton van Leeuwenhoek, the Dutch microscopist, began making and using the first microscope. With his microscope Leeuwenhoek could see the blood capillaries for the first time. He could also see the blood cells inside these capillaries. Harvey's ideas were proved to be correct.

## Measurement in chemistry

It was not until chemists began measuring the amounts of substances taking part in chemical reactions that they began to develop any clear ideas about the ways atoms are built into molecules, and molecules react one with the other. One of the pioneers in this field was Joseph Black. In the 18th century he studied the production of quicklime from chalk, the production of slaked lime from quicklime and several other reactions. He measured the weights of each substance taking part in the reactions, and the weights of each substance produced. He was then able to deduce the chemical composition of the substances involved.

Since the days of Black, measurement has been an essential part of the study of chemistry. Measurement has even led to the discovery of a new chemical element. The element was first discovered in 1868 by the French astronomer, Pierre Janssen. He was examining the spectrum of the Sun during an eclipse and noted a bright yellow line. During an eclipse the light we receive is coming direct from the outermost layers of the Sun, and so the spectrum appears as a series of bright coloured lines. Each line can be identified with the element that produces it, for we can measure the position of each line and compare it with the position of lines produced by heating known elements in the laboratory. Here was a line that could not be identified, yet it was bright, showing that this unidentified element was present in the outer layers of the Sun in noticeable amounts. The element was given the name *helium* after the Greek word *helios*, meaning Sun. Measurement of spectral lines had led to the discovery of helium in the Sun, before it had been found on Earth. Helium is rare on Earth and it is the more difficult to find because it is an *inert* substance. Inert substances are those that do not readily take part in chemical actions. This being so, it is difficult to test for them chemically. The search for helium began and it was thirteen years later when helium was detected in gases coming from the crater of Mount Vesuvius. Later, the gas was found in rocks and in the air. Today the major sources of helium are natural gas wells in the United States. It has very low density and, being inert, is incombustible, so it is often used in balloons and lighter-than-air aircraft.

## Science and the technology of measurement

The main purpose of this book is to show how the ability to measure has been so essential in the discovery of scientific knowledge. In short we can say: measurement helps science. We can also say the opposite with equal truth: science helps measurement. Scientific knowledge of the properties of materials and the way things behave, helps us to design and make better measuring instruments. Knowledge of the composition of glass has helped us make better lenses, so we now have better telescopes, microscopes, spectrometers and all manner of optical measuring instruments. Knowledge of the laws of motion, especially the motion of the pendulum, helped in the design of accurate clocks. Then it was noticed that changes in temperature caused the pendulum to alter in length and so the clock became inaccurate. Scientific knowledge of the expansion of different metals helped John Harrison (p.110) and others design pendulums that were not affected by temperature.

One of the most recent and most far-reaching ways in which science has helped measurement has been in the field of solid-state physics. This is the study of the structure and physical properties of crystalline materials. A large part of this has been the study of semiconducting materials. This lead to the invention of the transistor, by the Americans William Schockley, W. H. Brattain and

John Bardeen in 1946. Although electronics had been a flourishing technology ever since the invention of the thermionic valve by the Englishman, John Ambrose Fleming in 1904, the transistor was a major development. This and the other solid-state electronics devices that followed it have made vast differences to everyday life and to the progress of science. Now we can build instruments and machines that would have been almost impossible to build before the transistor arrived. Valves were large and expensive, they had a short life and required considerable amounts of electrical power to run them. Transistors are small, cheap, reliable and require very little power. The integrated circuit in a cheap digital watch contains the equivalent of a thousand or so transistors yet operates with high precision and accuracy for months on end, using only two miniature low-voltage dry cells.

From the atomic clock to the television cameras and other instruments landed by *Vikings* I and II on the surface of Mars, electronics has revolutionised our methods of measuring. This is the result of our researches into solid-state physics. Science and measurement each help the other. Together they led to an even better understanding of the nature of our Universe.

# Glossary

Alpha particle  An atomic particle consisting of two protons and two neutrons; the equivalent of the nucleus of a helium atom.

Ceramic  A material made by heating clay strongly, as when pottery is made.

Coulomb  The unit of electrical charge. If a current of 1 ampere flows past a given point in a circuit for a period of 1 second, the total amount of charge carried past that point is 1 coulomb. Symbol, C.

Disc  When a star or planet is observed from a great distance, it is not possible to see that it is a sphere. It appears as a circular (or nearly circular) disc. At even greater distances the disc can not be seen, the body appearing as a point of light.

Electromagnetic radiation  A type of radiation such as light, x-rays, radio waves and gamma rays, which is spread by oscillating electrical and magnetic fields.

Latitude  Lines of latitude are drawn on maps or globes of the Earth or other astronomical bodies for measuring (in degrees) distances north or south of the equator.

Longitude  Lines of longitude are drawn on maps or globes of the Earth or other astronomical bodies for measuring (in degrees) distances east or west of one particular line of longitude. The distance is usually measured from the line of longitude which passes through Greenwich, England, known as the Greenwich Meridian.

Mass    The amount of matter of which a body is composed. In SI, mass is measured in kilograms.

Power    The rate of doing work, or the rate of converting one form of energy into another form of energy. The unit is the watt, a rate of conversion of 1 joule per second.

Quartz    A mineral substance consisting of silica in crystalline form.

Ultra-sound    Sound waves of frequency which are too high to be heard by the human ear, though certain animals such as dogs, cats, bats and many others are able to hear it.

# Index